A Guide to Visual Multi-Level Interface Design From Synthesis of Empirical Study Evidence

Synthesis Lectures on Visualization

Editor

David S. Ebert, *Purdue University*

Synthesis Lectures on Visualization will publish 50- to 100-page publications on topics pertaining to scientific visualization, information visualization, and visual analytics. The scope will largely follow the purview of premier information and computer science conferences and journals, such as IEEE Visualization, IEEE Information Visualization, IEEE VAST, ACM SIGGRAPH, IEEE Transactions on Visualization and Computer Graphics, and ACM Transactions on Graphics. Potential topics include, but are not limited to: scientific, information, and medical visualization; visual analytics, applications of visualization and analysis; mathematical foundations of visualization and analytics; interaction, cognition, and perception related to visualization and analytics; data integration, analysis, and visualization; new applications of visualization and analysis; knowledge discovery management and representation; systems, and evaluation; distributed and collaborative visualization and analysis.

A Guide to Visual Multi-Level Interface Design From Synthesis of Empirical Study Evidence
Heidi Lam and Tamara Munzner
2010

A Guide to Visual Multi-Level Interface Design From Synthesis of Empirical Study Evidence

Heidi Lam and Tamara Munzner

ISBN: 978-3-031-01470-3   paperback
ISBN: 978-3-031-02598-3   ebook

DOI 10.1007/978-3-031-02598-3

A Publication in the Springer series
SYNTHESIS LECTURES ON VISUALIZATION

Lecture #1
Series Editor: David S. Ebert, *Purdue University*
Series ISSN
Synthesis Lectures on Visualization
ISSN pending.

A Guide to Visual Multi-Level Interface Design From Synthesis of Empirical Study Evidence

Heidi Lam
Google Inc.

Tamara Munzner
University of British Columbia

SYNTHESIS LECTURES ON VISUALIZATION #1

ABSTRACT

Displaying multiple levels of data visually has been proposed to address the challenge of limited screen space. Although many previous empirical studies have addressed different aspects of this question, the information visualization research community does not currently have a clearly articulated consensus on how, when, or even if displaying data at multiple levels is effective. To shed more light on this complex topic, we conducted a systematic review of 22 existing multi-level interface studies to extract high-level design guidelines. To facilitate discussion, we cast our analysis findings into a four-point decision tree: (1) When are multi-level displays useful? (2) What should the higher visual levels display? (3) Should the different visual levels be displayed simultaneously, or one at a time? (4) Should the visual levels be embedded in a single display, or separated into multiple displays?

Our analysis resulted in three design guidelines: (1) the number of levels in display and data should match; (2) high visual levels should only display task-relevant information; (3) simultaneous display, rather than temporal switching, is suitable for tasks with multi-level answers.

KEYWORDS

focus and context, overview and detail, zoomable user interfaces, fisheye view, multi-level display, empirical study, systematic review

Contents

Acknowledgments

We appreciate many discussions with Ted Kirkpatrick on paper structure, and we thank François Guimbretière, Diane Tang, Joanna McGrenere, and John Dill for their feedback on drafts. We also appreciate the suggestions of several anonymous reviewers.

Heidi Lam and Tamara Munzner
November 2010

Figure Credits

Figure 2 is from Lam, H., Munzner, T., and Kincaid, R. 2007. Overview Use in Multiple Visual Information Resolution Interfaces. IEEE Transactions on Visualization and Computer Graphics (TVCG) 13, 6,1278–1283. DOI: 10.1109/TVCG.2007.70583. © IEEE. Used with permission.

Figure 4 is from Hornbæk, K. and Frokjær, E. 2001. Reading of Electronic Documents: The Usability of Linear, Fisheye and Overview+Detail Interfaces. In Proc. ACM SIGCHI Conf. on Human Factors in Computing Systems (CHI'01). 293–300. DOI: http://dx.doi.org/10.1145/365024.365118. Copyright © 2001, Association for Computing Machinery, Inc. Reprinted by permission.

Figure A1 is from Baudisch, P., Good, N., Bellotti, V., and Schraedley, P. 2002. Keeping Things in Context: A Comparative Evaluation of Focus Plus Context Screens, Overviews, and Zooming. In Proc. ACM SIGCHI Conf. on Human Factors in Computing Systems (CHI'02). 259–266. DOI: 10.1145/503376.503423. Copyright © 2002, Association for Computing Machinery, Inc. Reprinted by permission.

Figure A2 is from Baudisch, P., Lee, B., and Hanna, L. 2004. Fishnet, a Fisheye Web Browser with Search Term Popouts: A Comparative Evaluation with Overview and Linear View. In Proc. ACM Advanced Visual Interfaces (AVI'04). 133–140. DOI: 10.1145/989863.989883. Copyright © 2004, Association for Computing Machinery, Inc. Reprinted by permission.

Figure A3 is from Bederson, B., Clamage, A., Czerwinski, M., and Robertson, G. 2004. Date-Lens: A Fisheye Calendar Interface for PDAs. ACM Trans. on Computer-Human Interaction (TOCHI) 11, 1 (Mar.), 90–119. DOI: 10.1145/972648.972652. Copyright © 2004, Association for Computing Machinery, Inc. Reprinted by permission.

Figure A4 is from Buring, T., Gerken, J., and Reiterer, H. 2006b. User Interaction with Scatterplots On Small Screens: A Comparative Evaluation of Geometric-Semantic Zoom and Fisheye Distortion. IEEE Trans. on Visualization and Computer Graphics (TVCG) 12, 5, 829–836. DOI:10.1109/TVCG.2006.187. © IEEE. Used with permission.

Figure A5 is from Cockburn, A. and Smith, M. 2003. Hidden Messages: Evaluating the Effectiveness of Code Elision in Program Navigation. Interacting with Computers, 15, 3, 387–407. © 2003 Elsevier. Used with permission.

Figure A13 is from Lam, H., Munzner, T., and Kincaid, R. 2007. Overview Use in Multiple Visual Information Resolution Interfaces. IEEE Transactions on Visualization and Computer Graphics (TVCG) 13, 6, 1278–1283. 10.1109/TVCG.2007.70583. © IEEE. Used with Permission.

Figure A14 is from Nekrasovski, D., Bodnar, A., McGrenere, J., Munzner, T., and Guimbretiére, F. 2006. An Evaluation of Pan and Zoom and Rubber Sheet Navigation. In Proc. ACM SIGCHI Conf. on Human Factors in Computing Systems (CHI'06). 11–20. DOI: 10.1145/1124772.1124830. Copyright © 2006, Association for Computing Machinery, Inc. Reprinted by permission.

Figure A15 is from North, C. and Shneiderman, B. 2000. Snap-Together Visualization: Can Users Construct and Operate Coordinated Visualizations. Intl. J. of Human-Computer Studies (IJHCS) 53, 5, 715–739. © 2000 Elsevier. Used with permission.

Figure A16 is from Pirolli, P., Card, S., and van der Wege, M. 2003. The Effects of Information Scent on Visual Search in the Hyperbolic Tree Browser. ACM Trans. on Computer-Human Interaction (TOCHI) 10, 1, 20–53. DOI: 10.1145/606658.606660. Copyright © 2003, Association for Computing Machinery, Inc. Reprinted by permission.

Figure A17 is from Plaisant, C., Grosjean, J., and Bederson, B. 2002. SpaceTree: Supporting Exploration in Large Node Link Tree, Design Evolution and Empirical Evaluation. In Proc. IEEE Symposium on Information Visualization (InfoVis'02). 57–64. © 2002 IEEE. Used with permission.

Figure A18 is from Plumlee, M. and Ware, C. 2006. Zooming versus Multiple Window Interfaces: Cognitive Costs of Visual Comparisons. Proc. ACM Transactions on Computer-Human Interaction (TOCHI) 13, 2, 179–209. DOI: 10.1145/1165734.1165736. Copyright © 2006, Association for Computing Machinery, Inc. Reprinted by permission.

Figure A19 is from Saraiya, P., Lee, P., and North, C. 2005. Visualization of Graphs with Associated Timeseries Data. In Proc. IEEE Symposium on Information Visualization (InfoVis'05). 225–232. © IEEE. Used with permission.

Figure A20 is from Schafer, W. and Bowman, D. A. 2003. A Comparison of Traditional and Fisheye Radar View Techniques for Spatial Collaboration. In Proc. Conf. on Graphics Interface (GI'03). 39–46. Copyright © 2003, Canadian Information Processing Society. Used with permission.

Figure A21 is from Schaffer, D., Zuo, Z., Greenberg, S., Bartram, L., Dill, J., Dubs, S., and Roseman, M. 1996. Navigating Hierarchically Clustered Networks through Fisheye and Full-Zoom Methods. ACM Trans. on Computer-Human Interaction (TOCHI) 3, 2 (Mar.), 162–188.

CHAPTER 1

Introduction

Visualization designers often need to display large amounts of data that exceed the display capacity of the output devices, and arguably, the perceptual capacity of the users. Displaying data at multiple visual levels has been suggested as a workaround for this design challenge. Examples of interfaces with multiple visual levels include zooming, focus + context, and overview + detail interfaces.

Even though it is generally believed that visualization interfaces should provide more than one visual level (e.g., (p. 307) Card et al. [1999]), we as a community still face considerable uncertainty as to when and how multi-level interfaces are effective, despite numerous evaluation efforts Cockburn et al. [2008], Furnas [2006]. One challenge is that many previous studies and reviews compare between visualization interfaces at a coarse-grained, monolithic level. While these studies provide holistic insight on whether a particular interface works for a particular task and users, visualization designers have difficulty in directly using these study results for guidance as they design new interfaces.

To make more informed decisions, we need to look beyond the entire interface to tease out the factors at play that significantly affect its use. These factors include the *interface element* factor of the type and amount of information displayed, how it is visually transformed, and what interactions are supported; the *task* factor of what information the intended use requires; and the *data* factor of how the data is intrinsically organized, such as whether it has hierarchical structure. In this paper, we carried out a fine-grained analysis of previous studies to identify the factors relevant for the design of multi-level visualizations, and further characterize their interplay. In most cases, we examine how the interaction of the task and the data factors affect the choice of interface elements.

We analyzed 22 existing multi-level interface studies to obtain a clearer snapshot of the current understanding of multi-level interface use, and how to apply this knowledge in their design. To unify our discussion, we grouped the interfaces into single or multi-level interfaces. For single-level interfaces, we looked at the *loLevel* interface that shows data in the highest available detail, for example, the "detail" in overview + detail interfaces. We considered three multi-level interface types in this review: *temporal*, or temporal switching of the different levels as in zooming interfaces; *separate*, or displaying the different levels simultaneously but in separate windows as in overview + detail interfaces; and *embedded*, or showing the different levels in a unified view as in focus + context interfaces. Chapter 2 includes a more detailed explanation of our terminology.

One difficulty of fine-grained analysis is to present the complex picture of the findings in a comprehensible way. We structure our discussion in terms of a decision tree for designers, as shown in Figure 1.1.

Considerations

5.1   Multi-level interaction costs
5.2   Suitability of single-level data

6.1   The number of visual levels
6.2   The amount of information
6.3   Information perceivability
6.4   A priori automatic filtering

7.1   Tasks with multi-level answer
7.2   Tasks with multi-level information
       clue
7.3   Tasks with single-level answer and
       information clue

8.1   Distortion

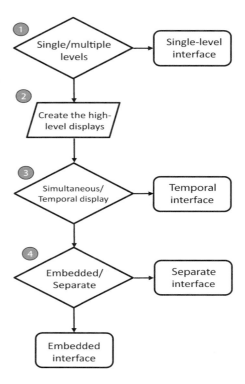

Figure 1.1: Decision tree to create a multi-level display. There are four major steps in the decision process, each covered in a chapter in the paper: (Decision 1/Chapter 5) Decide if a multi-level display is appropriate for the application; (Decision 2/Chapter 6) Decide on the number of levels, amount of data, and visual information to be displayed on the high-level displays; (Decision 3/Chapter 7) Decide on the methods to display the multiple levels; (Decision 4/Chapter 8) Decide on the spatial layout of the multiple visual levels. Considerations at each decision point are listed with their respective chapter numbers.

Our decision tree has four major steps: (1) Single- or multi-level interface; (2) Create the high-level displays; (3) Simultaneous or temporal display of the visual levels; and (4) Embedded or separate display of the visual levels. We now summarize design issues we examined at each of these steps based on empirical evidence extracted from papers we analyzed.

DECISION 1 (Chapter 5): Single- or multi-level interface The first step in the process is to decide if a multi-level interface is suitable for the task and data at hand. Even though multi-level interfaces allow more flexible displays of data, the choice is not obvious as multi-level interfaces typically have more complex and involved interactions than their single-level counterparts. The design needs to balance interaction costs and display flexibility. Chapter 5.1 discusses interaction costs reported in the reviewed studies. Chapter 5.2 discusses considerations in using multiple levels to display single-level data.

DECISION 2 (Chapter 6): Create the high-level displays If the designer decides to use a multi-level interface for the data, the next step in the design process is to create the high-level displays, which is a challenge with large amounts of data Keim et al. [2006]. In addition to the technical challenges in providing adequate interaction speed and in fitting the data onto the display device, the designer also needs to consider the appropriate number of levels of visual information provided by the interface. Study results indicate that providing too many visual levels may be distracting to users, as discussed in Chapter 6.1. Similarly, showing too much data in the high-level displays can also be distracting, as discussed in Chapter 6.2. In many cases, the data may have to be abstracted and visually abbreviated to increase the display capability of the high-level displays. Ellis and Dix [2007] provides a taxonomy of clutter reduction techniques that include sampling, filtering, and clustering. Chapter 6.3 discusses cases where designers have gone too far in their abstractions and study participants could no longer use the visual information on the high-level displays. Instead of abstraction, the designer could choose to selectively display or emphasize a subset of the data on the high-level displays, for example, based on the generalized fisheye degree-of-interest function Furnas [1986]. However, study results suggest that *a priori* automatic filtering may be a double-edged sword, as discussed in Chapter 6.4. Given all these considerations, we complete the discussion by re-examining the roles of high-level displays in Chapter 6.5 to help ground high-level display design.

DECISION 3 (Chapter 7): Simultaneous or temporal display of the visual levels Once the visual levels are created, the designer then needs to display them, either simultaneously as in the *embedded* or the *separate* interfaces, or one visual level at a time as in the *temporal* interfaces. Generally, *temporal* displays require view integration over time and can therefore burden short-term memory Furnas [2006]. On the other hand, simultaneous-level interfaces have more complex interactions such as view coordination in *separate* displays and the issue of image distortion frequently found in *embedded* displays. Our reviewed studies found that simultaneous multi-level interfaces were beneficial for tasks that have multi-level answers or multi-level clues to single-level answers. Otherwise, *temporal* displays may be more suitable. Chapters 7.1 and 7.2 consider the multi-level answer and multi-level clues cases respectively, while Chapter 7.3 examines the single-answer, single-clue case.

DECISION 4 (Chapter 8): Embedded or separate display of the visual levels If the choice is simultaneous display of the multiple visual levels, the designer then has to consider the spatial layout of the levels. The choices are to display the visual levels in the same view, as in the *embedded* interfaces, or by showing them in separate views, as in the *separate* interfaces. Both spatial layouts involve trade offs: the *embedded* displays frequently involve distortion, as discussed in Chapter 8.1, and the *separate* displays involve view coordination. We discuss each of these decision points starting from Chapter 5. For each chapter, we summarized current beliefs and assumptions about multi-level interface use, along with relevant study results. We also flagged situations where study results did not clearly support our previous beliefs based on the existing literature.

In summary, the contribution of this paper is a set of evidence-based guidelines for the design of multi-level visualization interfaces, presented in the form of a decision tree. We believe that we are the first to do so through the fine-grained analysis of previous studies at the level of factors. Design decisions based on coarse-grained analysis of previous studies at the level of the whole interface may not be accurate due to imperfect matching of factors such as tasks or the number of levels in data. While it is useful to summarize existing study results at the study level to get an overall sense of our communal knowledge on these interfaces (e.g., Cockburn et al. [2008]), it is important to look at the same set of experiments at a finer-grained level to inform specific designs.

CHAPTER 2

Terminology

Due to the diverse nature of interfaces examined in this paper, we use the term **visual level** as a general measure of visual information displayed. Visual level encompasses three measures: data hierarchy, visual quantity, and visual quality of the displayed data:

1. **Data hierarchy**: Higher visual levels show data at higher levels of the data hierarchy. For example, in treemaps, users can focus on different layers of the hierarchical tree at different levels in the display Bederson et al. [2002].

2. **Visual quantity**: Higher visual levels display less details. One example is semantic zooming, where users are provided with different amounts of detail in a visual level by zooming in and out. This is akin to the "levels-of-detail" concept, as in low-level details.

3. **Visual quality**: Higher visual levels display data objects with less perceivable encodings. One common example is the display of textual data. With the same font type, data displayed using small unreadable font sizes is considered to be at a higher visual level than those displayed in larger readable font sizes. One example is displaying greeked text as the high-level display in document readers (e.g., Figure 2.3, the Fisheye interface Hornbæk and Frokjær [2001]). As for visual objects, the criteria of perceivability is less well defined. One example is the visual encodings used in one of our reviewed studies. In Lam et al. [2007], the same line graph data was encoded using two different types of encodings (Figure 2.1). The low-level visual encoding displays the y-dimension of line graphs using both space and colour, while the high-level encoding only uses colour, thus making the fine details of the displayed line graph less perceivable. This concept is akin to "scale", where the zoomed-out view is of higher visual level than the zoomed-in view.

Several taxonomies of multi-level visualization techniques exist, for example the taxonomy for image browsers in Plaisant et al. [1995]. While many of the previous taxonomies focus on expected functions of the interfaces, we instead focus on the visual encodings: for example, focus (as in focus + context) or detail (as in overview + detail) can be thought of as a **low visual level**, while context or overview is of comparatively **high visual level**. Multi-level interfaces can be further classified as **temporal** or **simultaneous** based on the way they display the visual levels, as shown in Figure 2.2. **Temporal** interfaces, an example being pan-and-zoom user interfaces, allow users to drill up and down the zoom hierarchy and display the different visual levels one at a time. In contrast, **simultaneous** interfaces show all the visual levels on the same display. We refer to interfaces that integrate and spatially embed the different levels as **embedded** displays, as in focus + context visualizations. When

Figure 2.1: Two visual encodings of the same line-graph data. Despite showing the same amount of data points, the two visual encoding are of different visual levels. (a) LoLevel: encodes the y-dimension of the line graph with both space and colour; (b) HiLevel: encodes the y-dimension with only colour.

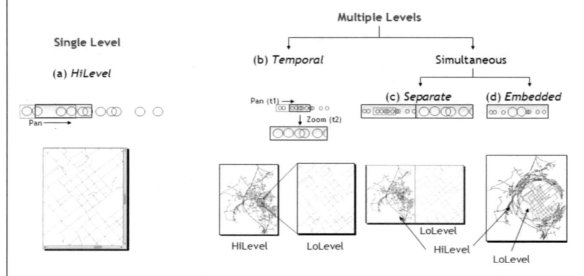

Figure 2.2: Our classification of interfaces. Interfaces can be classified based on the number of visual levels. (a) Single-level interfaces display one visual level, as in panning interfaces. (b)–(d) multi-level interfaces show multiple visual levels. In this illustration, each interface contains two visual levels: low (denoted as large circles) and high (denoted as small circles). (b) In the *temporal* approach, users can pan around in the high-level view and zoom into an subarea as a low-level view, as in pan and zoom interfaces. (c) In the *separate* approach, the high and the low levels can be placed in separate panels, as in overview + detail interfaces. (d) In the *embedded* approach, the low and the high visual levels are embedded in a single unified display, as in focus + context interfaces.

the different visual levels are displayed as separate views, we refer to these interfaces as **separate**, as in overview + detail displays. Since the different levels can occupy the entire display window, or be integrated as part of a single window, we explicitly differentiate the two by using the term **view** to denote individual window or pane, and the term **region** to denote an area within a view. This classification into temporal, simultaneous, and embedded matches three of the categories recently proposed by Cockburn et al. [2008], but we use different terminology to emphasize the relationships between these categories, rather than their traditional names. We differ by not including their *cue* category, because it crosscuts the other three and thus does not directly address our core concern of multiple visual levels. We now use one of our reviewed studies to illustrate our terminology. Screen captures of all reviewed interfaces can be found in Appendix A. In the study reported in "Reading Patterns and Usability in Visualization of Electronic Documents" by Hornbæk and Frokjær [2001], the study included three interfaces to display text documents, as shown in Figure 2.3:

1. A linear interface that was vertically scrollable and displayed text in normal fonts (Figure 2.3(a)). We classify this interface as *loLevel*;

2. An overview + detail interface that showed a 1:17 reduction of the text document as an overview, with section and subsection headers of the document being readable (Figure 2.3(b)). The original document constituted the detailed view. We have classified this interface as *separate*.

3. A fisheye interface that showed all except the most important parts of the document in reduced (and unreadable) font in order to fit the document on the display (Figure 2.3(c)). We have classified this interface as *embedded*.

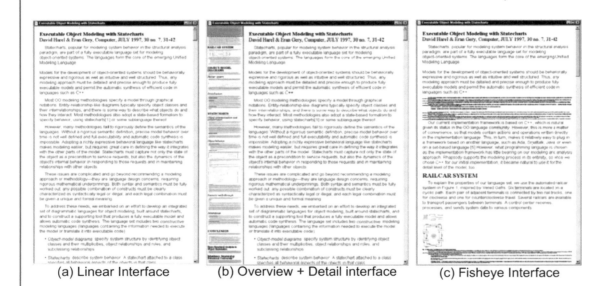

(a) Linear Interface   (b) Overview + Detail interface   (c) Fisheye Interface

Figure 2.3: Concrete examples of our interface types. Interfaces from Hornbæk et al. [2001] are used to illustrate our interface classification system. (a) The linear interface is classified as *loLevel*; (b) The overview + detail interface is classified as *separate*; (c) The fisheye interface is classified as *embedded*.

CHAPTER 3

Methodology

In this chapter, we explain our study approach, our paper selection criteria, and our result presentation method. To contrast our approach with existing work, we discuss the related work here.

3.1 STUDY APPROACH

Ideally, we would like to perform a meta-analysis on our reviewed studies. In other words, we would like to be able to translate study results from different studies to common metrics and statistically explore relationships between study characteristics and study results. However, recognizing that our reviewed studies are different in their implementations of visualization techniques, study tasks and data, and in some cases, experimental design and measurements, we may only be able to include a very small subset of existing studies in such an analysis. Indeed, only 6 of the 35 studies considered in Chen and Yu [2000] met their meta-analysis criteria. Part of the difficulty in conducting systematic reviews based on existing studies is the lack of standards in both study design and reporting Lam and Munzner [2008]. Some of Chen and Yu [2000]'s recommendations, echoed by others (e.g., Ellis and Dix [2006], Plaisant [2004]), are still active areas of research. One example is to create standardized task taxonomies for interface evaluations (e.g., Valiati et al. [2006], Winckler et al. [2004]).

In order to make progress before our community reaches a consensus on visualization evaluations, we took a qualitative bottom-up approach. Instead of answering specific questions (e.g., Chen and Yu [2000], Hudhausen et al. [2002]), we aimed to discover emergent themes from existing study results.

Our goals are therefore to understand multi-level interface use and to extract design guidelines. Our process is illustrated in Figure 3.1. We started our process by first collecting eligible existing studies, as discussed in the next section. We then coded identified studies based on the interfaces studied, as shown in Table 4.1, and by major study results, as shown in Appendix A. For study results, we focused on objective measures of task time and accuracy since these measures were reported in all reviewed studies. To interpret these results, we also considered explanations provided by paper authors based on their observations and understanding of their studies.

Due to the diverse and disparate nature of the studies reviewed, we did not directly compare study results between studies. Instead, we compared results within each study by pairing up interfaces (e.g., *embedded* and *separate*). To identify possible reasons that may explain study results, we considered three interface factors:

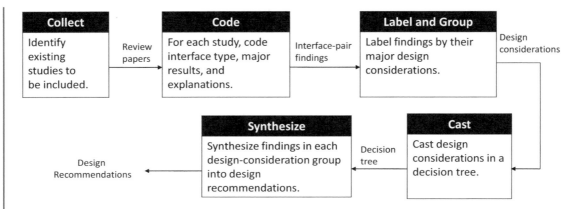

Figure 3.1: Analysis process. We took a qualitative and bottom-up approach in our analysis to obtain design recommendations for multi-level interface design. Our process has five steps: collect studies for coding, code studies for interface-pair findings, label and group findings to extract design considerations, cast considerations into a decision tree, and synthesize considerations at each step of the tree to obtain design recommendations.

1. **Interface elements** as in visual encoding and transformations, number and spatial arrangements of visual levels, and supported interactions

2. **Task** as in task information requirements

3. **Data** as in intrinsic organization, e.g., hierarchical structured

Since most of the existing multi-level interface studies did not explicitly consider user characteristics such as visual-spatial ability, we did not address this important issue in our discussion.

The result of this step in the process was a list of **interface-pair findings**, each consisted of interface pair, study results, and possible explanations of these results. Here is one example from [INFOSCENT] Pirolli et al. [2003]:

- *Interface pair*: *T* (*temporal*), *E* (*embedded*)
- *Study results*: Time: $T < E$ (low-scent tasks); $T < E$ (high-scent tasks);
- *Interface factors*: Data/Task: Data-requirement of tasks (low scent vs. high scent); Interface: spatial arrangement of visual levels;
- *Reasons*: Low-scent tasks did not provide clues at all visual levels so simultaneous display of multiple visual levels did not provide enough benefits. In contrast, high-scent tasks provided clues at multiple visual levels. Being able to see them all may have facilitated answer searching.

These interface-pair findings were then labeled with a **design consideration** based on the interface factors considered. In the [INFOSCENT] example, the design consideration was "How did distribution of task clues affect the choice of simultaneous or temporal display of the visual levels?". In this and most cases, the design consideration involves how the interaction of the task and the data factors affect the choice of interface elements.

To facilitate discussion, we organized these design considerations into a **decision tree** to construct a multi-level interface (Figure 1.1). The results were **design recommendations**. The [INFOSCENT] design consideration was slotted in "Decision 3: Simultaneous or temporal display of the visual levels", where we concluded that the *temporal* interface would be a better choice for single-level answers with single-level clues over the *embedded* or the *separate* interfaces, as discussed in Chapter 7.

3.2   PAPER SELECTION

We collected an initial set of candidate papers by performing keyword searches on popular search engines (Google and Google Scholar) and large academic databases (ACM and IEEE digital libraries), along with our own collection of study publications accumulated over the years. We further located more study publications based on citations of the initial set. During the course of our synthesis, we continuously added new publications until September 2007.

Due to the need for pairwise comparison in our approach, we only included papers that:

1. Studied at least two of the four interface types analyzed in our study: *loLevel*, *temporal*, *embedded*, and *separate*;

2. Included 2D or 2-1/2D interfaces only, since introducing a third visual dimension involves considering an additional and different set of factors such as view projection and occlusion, which we consider to be beyond the scope of this study;

3. Displayed comparable data sets on the interfaces;

4. Studied comparable tasks.

3.3   STUDY RESULT PRESENTATION

The design considerations that we offer are based on conclusions we draw from our synthesis, supported by evidence from the reviewed studies. In order to facilitate discussion, we organized these considerations as a four-point decision tree as the framework of our review (Figure 1.1). Since many studies included more than two study interfaces, their study results are mentioned in more than one sections of the paper.

3.4   RELATED WORK

There have been a number of review papers on different aspects of multi-level interfaces. We discuss here the most relevant two, that include a significant amount of analysis and synthesis.

Leung and Apperly's early review on the use of distortion in focus + context interfaces incorporated an influential taxonomy and an analysis unifying the mathematical framework behind the previously published techniques, but of course does not cover the explosion of work that has appeared since its publication in 1994 Leung and Apperley [1994].

The recent review paper by Cockburn et al. is an excellent summary of the current state of the art Cockburn et al. [2008], including not only papers that present interface techniques but also empirical evaluation papers, and a thorough discussion of previous reviews. Their synthesis of the previous empirical evaluations is a discussion of technique strengths and weaknesses in terms of low-level versus high-level user tasks, culminating in a small set of guidelines at the *interface* "to identify effective and ineffective uses of [visualizations]" (p. 2).

In contrast, we analyze and synthesize experimental results in a more fine-grained approach at the *factor* level, and thus present a much more detailed set of guidelines that take into account the complex interplay between these factors. The limitations of our approach include basing our analysis on a smaller set of papers, as discussed in 3.2; we exclude results from some well-conducted evaluations due to our need to tease out effects of contributing factors. A more detailed discussion of limitations of our approach is in Chapter 9.

CHAPTER 4

Summary of Studies

Table 4.1 lists the 22 studies reviewed, along with our coded interface type: *loLevel*, *temporal*, *separate*, and *embedded*. Screen captures of study interfaces are available in Appendix A.

We considered all interfaces in the reviewed studies except for the Saraiya et al. [2005] study, where we compare a *loLevel* interface to a *temporal* one but leave out their two "Multiple View" interfaces that displayed the same data in separate views at the same level, but used a different graphical format. Since our review focused on multi-level interfaces, we considered the issue of multiple presentation forms to be beyond the scope of our review.

This study aims to provide an evidence-based guide to designers in using multi-level interfaces, rather than being a review paper on existing multi-level study results, so we only provide enough study details to illustrate our points. For reference, Appendix A provides brief summaries of each study and lists the interfaces, tasks, data, and significant results for each of the reviewed papers.

For each design consideration, we list studies included for the analysis. Each paper is designated with an identification tag which is used in subsequent tables. Tags are assigned based on major subjects of investigation in the papers:

- Names of novel interfaces or techniques (e.g., [FISHNET] Baudisch et al. [2004] and [DATELENS] Bederson et al. [2004]);

- Data types (e.g., [SCATTERPLOT] Buring et al. [2006b], [EDOC] Hornbæk and Frokjær [2001], Hornbæk et al. [2003], and [LINEGR] Lam et al. [2007]);

- Existing techniques and tasks (e.g., [ZUINAV] Hornbæk et al. [2002], [FISHSTEER] Gutwin and Skopik [2003]);

- Phenomena (e.g., [INFOSCENT] Pirolli et al. [2003] and [VISMEM] Plumlee and Ware [2006]);

Please note that Hornbæk et al.'s online document study was reported in two papers: Hornbæk and Frokjær [2001] and Hornbæk et al. [2003].

Table 4.1: A list of multi-level studies reviewed. An 'x' in the right columns denotes the study included an interface of the type: L = LoLevel; T = Temporal; E = Embedded; S = Separate.

Tag	Authors	Paper Title	Single	Multiple		
			L	T	E	S
[FCScreen]	Baudisch et al. [2002]	Keeping things in context: a comparative evaluation of focus plus context screens, overviews, and zooming		x	x	x
[Fishnet]	Baudisch et al. [2004]	Fishnet, a fisheye web browser with search term popouts: a comparative evaluation with overview and linear view	x		x	x
[DateLens]	Bederson et al. [2004]	DateLens: a fisheye calendar interface for PDAs		x	x	
[ScatterPlot]	Buring et al. [2006b]	User Interaction with Scatterplots On Small Screens—A Comparative Evaluation of Geometric-Semantic Zoom and Fisheye Distortion		x	x	
[ElideSrc]	Cockburn and Smith [2003]	Hidden messages: evaluating the efficiency of code elision in program navigation	x		x	
[FishSteer]	Gutwin and Skopik [2003]	Fisheyes are good for large steering tasks			x	x
[BigOnSmall]	Gutwin and Fedak [2004b]	Interacting with big interfaces on small screens: a comparison of fisheye, zoom, and panning techniques	x	x	x	
[eDoc]	Hornbæk and Frokjær [2001]	Reading of electronic documents: the usability of linear, fisheye and overview + detail interfaces	x		x	x
	Hornbæk et al. [2003]	Reading patterns and usability in visualization of electronic documents	x		x	x
[ZuiNav]	Hornbæk et al. [2002]	Navigation patterns and usability of zoomable user interfaces with and without an overview		x		x
[FishMenu]	Hornbæk and Hertzum [2007]	Untangling the usability of Fisheye menus		x	x	x
[FishSrc]	Jakobsen and Hornbæk [2006]	Evaluating a fisheye view of source code	x		x	
[SumThum]	Lam and Baudisch [2005]	Summary Thumbnails: readable overviews for small screen web browsers	x	x		
[LineGr]	Lam et al. [2007]	Overview use in multiple visual information resolution interfaces	x		x	x
[RubNav]	Nekrasovski et al. [2006]	An evaluation of pan and zoom and rubber sheet navigation		x	x	x
[Snap]	North and Shneiderman [2000]	Snap-together visualization: can users construct and operate coordinated visualizations?	x			x
[InfoScent]	Pirolli et al. [2003]	The effects of information scent on visual search in the hyperbolic tree browser		x	x	
[SpaceTree]	Plaisant et al. [2002]	SpaceTree: supporting exploration in large node link tree, design evolution and empirical evaluation		x	x	
[VisMem]	Plumlee and Ware [2006]	Zooming, multiple windows, and visual working memory		x		x
[TimeGr]	Saraiya et al. [2005]	Visualization of graphs with associated timeseries data	x	x		
[FishRadar]	Schafer and Bowman [2003]	A comparison of traditional and fisheye radar view techniques for spatial collaboration			x	x
[FishNav]	Schaffer et al. [1996]	Navigating hierarchically clustered networks through fisheye and full-zoom methods		x	x	
[SpaceFill]	Shi et al. [2005]	An evaluation of content browsing techniques for hierarchical space-filling visualizations		x	x	

CHAPTER 5

Decision 1: Single or Multi-level Interface?

The first step in our design decision tree is to decide if a multi-level interface is appropriate for the task and data at hand. To isolate situations where the additional high visual levels were found to be useful, we looked at studies that compared single-level *loLevel* interfaces to one of the three multi-level interfaces: *temporal*, *embedded*, and *separate*.

It is generally believed that interfaces should provide more than one visual level (e.g., (p. 307) Card et al. [1999]). However, for users, having extra visual levels means more complex and difficult level coordination and integration, which may be time consuming and require added mental and motor efforts (p. 393) Cockburn and Smith [2003]. Interaction costs may be justified if higher visual levels provided in addition to the basic *loLevel* displays are useful to users. This chapter examines the factors of interaction costs in multi-level interface (Chapter 5.1) and matching of data levels required by the task with the number of visual levels in the display. Study results suggest that single-level data may not be suited for multi-level display, as discussed in Chapter 5.2.

5.1 MULTI-LEVEL INTERFACE INTERACTION COSTS SHOULD BE CONSIDERED

Interaction complexity can be difficult to measure and isolate, but nonetheless may severely affect the usability of an interface Lam [2008]. Commonly used objective measurements such as performance time and accuracy are aggregate measures and cannot be used to identify specific interaction costs incurred in interface use. In seven of our reviewed papers, researchers recorded usage patterns, participant strategies, and interface choice that revealed interaction costs (Table 5.1).

5.1.1 INTERACTION COSTS FROM USAGE PATTERNS

As shown in Table 5.1, 5 of the 22 studies reported usage patterns constructed based on eye-tracking records or navigation action logs. Two of the studies reported usability problems with their multi-level interfaces Hornbæk et al. [2002], Hornbæk and Hertzum [2007].

[ZuiNav] studied map navigation using zoomable user interfaces with or without an overview Hornbæk et al. [2002]. Hornbæk et al. reported that participants who actively used the high-level view switched between the low and the high visual levels more frequently, which resulted in longer task completion time, as the additional high-level view may require mental effort and time

Table 5.1: Seven papers reported interface interactions, some with multiple categories. Five reported usage patterns obtained either from eye-tracking records or navigation-action logs; two reported participant strategies; and three reported interface choice.

Source	Papers
Usage patterns (Eye-tracking records)	[FISHMENU] Hornbæk and Hertzum [2007] [INFOSCENT] Pirolli et al. [2003]
Usage patterns (Navigation-action logs)	[EDOC] Hornbæk et al. [2003] [ZUINAV] Hornbæk et al. [2002] [FISHSRC] Jakobsen and Hornbæk [2006]
Participant strategies	[FCSCREEN] Baudisch et al. [2002] [LINEGR] Lam et al. [2007]
Interface choice	[EDOC] Hornbæk and Frokjær [2001]; Hornbæk et al. [2003] [ZUINAV] Hornbæk et al. [2002] [LINEGR] Lam et al. [2007]

moving the mouse, thus adding complexity in the interaction (p. 382). Indeed, navigation patterns showed that only 55% of the 320 tasks were solved with active use of the high-level view in their multi-level interfaces (p. 380).

[FISHMENU] studied item searching using fisheye menus Hornbæk and Hertzum [2007]. The paper reported large navigation costs in their *separate* and *embedded* interfaces, all implemented with a focus-locking interaction mechanism Bederson [2000]. Even though these interfaces succeeded in facilitating quick, coarse navigation to the target, participants had difficulty getting to the final target since the menu items moved with the mouse. Based on eye-tracking data, Hornbæk and Hertzum [2007] reported that participants had longer fixations and longer scan paths with their *separate* and *embedded* interfaces than with their *temporal* interface, suggesting increased mental activity and visual search.

5.1.2   INTERACTION COSTS FROM PARTICIPANT STRATEGIES

As shown in Table 5.1, 2 of the 22 studies reported participant strategies in interface use.

[FCSCREEN] studied map navigation, map path-finding, and verification using a *loLevel*, a *temporal*, and a *embedded* interface Baudisch et al. [2002]. The paper reported that some participants avoided continuously zooming in and out using the *temporal* interface by memorizing all the locations required in the task, and they answered the questions in a planned order. As a result, they could stay at a specific magnification without zooming back to the high-level view, thus effectively using the *temporal* interface as a *loLevel* interface.

In [LINEGR], participants developed a strategy to use the seemingly suboptimal *loLevel* interface in a visual comparison task Lam et al. [2007]. The study data consisted of a collection of line graphs that were identical except shifted by various amounts in the horizontal dimension. The task involved matching a line graph with the same amount of horizontal shift. Some participants took advantage of spatial arrangement of the *separate* interface by selecting candidate line graphs from the high-level view and displaying them at a low level for side-by-side comparison. The majority of participants, however, developed a strategy to enable the use of the low-level view alone. Taking advantage of the mouse wheel and the tool-tips, which displayed horizontal and vertical values of the line graph point under the cursor, participants scrolled vertically up and down with the cursor fixed horizontally at the point where the target peaked. As a result, they eliminated the need to visually compare line graphs. Instead, they tried to find another peak at the same x point numerically, by reading off the tool-tips, and avoided the need to interact with multiple visual levels.

5.1.3   INTERACTION COSTS FROM PARTICIPANTS' INTERFACE CHOICES

Another indicator of interaction costs is the active choice of the participants to use only one visual level in a multi-level interface to avoid coordinating between the multiple levels. As shown in Table 5.1, participants could explicitly convert a multi-level into a single-level interface in 2 of the 22 studies, and in [ZUINAV], Hornbæk et al. [2002] recorded active pane use on map navigation.

In [EDOC], a study on reading electronic documents, participants could expand all the document sections at once by selecting the pop-up menu item "expand all" in the *embedded* interface Hornbæk and Frokjær [2001], Hornbæk et al. [2003]. Six out of 20 participants chose to do this in one or more of the tasks. On average, they expanded 90% of the sections, thus effectively using the *embedded* interface as a *loLevel* interface.

In [ZUINAV], a map navigation study, 45% of participants did *not* actively use the high-level view in the *separate* interface even though 80% of participants reported preference for having the extra high-level view Hornbæk et al. [2002].

In [LINEGR], a study on visual search and comparison of line graphs, participants could expand all initially compressed graphs in their *embedded* or their *separate* interface by a key press, thus effectively turning the multi-level interface into a single-level interface Lam et al. [2007]. Their participants actively switched to the *loLevel* interface in 58% of the trials.

We suspect this desire to use only a single visual level when given a multi-level interface is more prevalent than reported. In many cases, participants were not provided with a simple mechanism to convert from the multi-level interface to its single-level counterparts, while in other cases, sole use of one window in the *separate* interface could not be discerned without detailed interaction recordings such as eye-tracking records. Using multi-level interfaces as single-level interfaces may explain the inability of some studies to distinguish *loLevel* interface and their multi-level counterparts.

5.2   SINGLE-LEVEL TASK-RELEVANT DATA MAY NOT BE SUITED FOR MULTI-LEVEL DISPLAYS

Table 5.2: Seven papers had single-level data and included a single-level interface for analysis comparison. In these cases, most multi-level interfaces supported the same or worse performance than their single *loLevel* counterparts.

Multi-level Effect	Paper with single-level data
No benefits	[FISHNET] Baudisch et al. [2004]
	[ELIDESRC] Cockburn and Smith [2003]
	[LINEGR] Lam et al. [2007]
Adverse effects	[ZUINAV] Hornbæk et al. [2002]
Mixed effects	[BIGONSMALL] Gutwin and Fedak [2004b]
	[EDOC]Hornbæk et al. [2003]
Excluded	[SUMTHUM] Lam and Baudisch [2005]

In this section, we examine how the data/task factors affect the choice of interface visual levels. Study results suggest that the number of visual levels provided by the interface should reflect the levels of organization in the data required by the task. Otherwise, users may need to pay the cost of coordinating between different visual levels without the benefits of rich information at every level. Among the seven studies reviewed that included a single-level interface, five of them used at least one set of single-level data (Table 5.2). Two studies failed to show performance benefits of multi-level interface for single-level data in cases where the tasks required detailed information not provided by the low-level display alone. [ZUINAV] showed adverse effects in using multi-level interfaces for single-level data Hornbæk et al. [2002]. [EDOC], a study on online documents, showed mixed results, as task nature affected the levels of data required, and consequently, interface use Hornbæk and Frokjær [2001], Hornbæk et al. [2003]. [BIGONSMALL] also showed mixed effects as one of their multi-level interfaces, their *embedded* Fisheye, performed well for single-level data but not their *separate* Two-level zoom interface Gutwin and Fedak [2004b] . We excluded [SUMTHUM] in this discussion as their *loLevel* interface had almost nine times the number of pixels than their multi-level interfaces Lam and Baudisch [2005], making direct comparisons difficult.

In order to understand effects of data level on interface effectiveness, we first needed to infer the amount of information required by study tasks based on task nature. For this consideration, we isolated situations where information required by the tasks was available at a single data level and compared participant performance and preference between single- and multi-level interfaces. In cases where data could be found on high-level displays, only the high-level displays in multi-level interfaces may be actively used by the participants, thus resulting in misleading judgements about the effectiveness of multi-level interfaces. We therefore contrasted such cases with those with that required information from low-level displays.

5.2.1   MULTI-LEVEL INTERFACES SHOWED NO BENEFITS

[ELIDESRC] examined the efficiency of program code elision in code navigation using a *loLevel* Flat text interface and two *embedded* elision interfaces Cockburn and Smith [2003]. Participants were asked to perform four tasks, of which two required single-level information: the *Signature retrieval* task required finding method argument types that were concentrated in the high-level displays; the *Body retrieval* task required finding the first call of a specific method, and thus required detailed information found only in the low-level displays. One of their *embedded* elision interfaces, the Illegible elision, showed benefits in the *Signature retrieval* task as the needed information were not elided and effectively concentrated on the high-level display of the *embedded* Illegible elision interface. In contrast, when the information is only found in the low-level displays, as in their *Body retrieval* task, none of the multi-level interfaces demonstrated any performance benefits, probably due to "the cost of configuring the level of elision when reading the method contents" (p. 402).

The situation is similar in [FISHNET], a study on information searches on web documents Baudisch et al. [2004]. These documents were displayed with guaranteed legible keywords, which constituted the high-level displays in the study's two multi-level interfaces: *embedded* Fishnet and *separate* overview. When the task only required reading the keywords, as in their *Outdated* task, their multi-level interfaces outperformed their *loLevel* browser, probably because the high-level displays in their multi-level interfaces concentrated task-relevant information in smaller display spaces, and we suspected their participants only used the high-level display in their multi-level interfaces. In contrast, when the task required reading surrounding text which may be too small to be legible in the high-level displays, as in the *Analysis* task, their multi-level interfaces did not support better participant performance. In short, when study tasks required information on the low-level displays, having a high-level display (i.e., having a multi-level display) did not result in performance benefits.

A similar situation is found in [LINEGR], a study that examined visual-target search on line-graph collections Lam et al. [2007]. Their multi-level interfaces (*embedded* and *separate*) only showed performance benefits over their *loLevel* scrolling interface when the visual targets could be directly identified on the high-level display, for example, in their *Max* task. Otherwise, having the extra high visual level did not seem to enhance participant performance.

5.2.2   MULTI-LEVEL INTERFACES SHOWED ADVERSE EFFECTS

[ZUINAV], a study on map navigation, reported adverse effects of displaying single-level data using a multi-level interface Hornbæk et al. [2002]. Despite having a similar number of objects, area occupied by the geographical state object, and information density on the maps, there were surprisingly large differences in usability and navigation patterns between trials using different data maps. The Washington-map trials had better performance time, accuracy and subjective satisfaction than the Montana-map trials. Hornbæk et al. [2002] explained these differences by differences in map content and the number of organization levels: the Washington map had three levels of county, city, and landmark, while the Montana map was single-leveled with weak navigation cues at low zoom levels.

5.2.3   MULTI-LEVEL INTERFACES SHOWED MIXED EFFECTS

Two studies showed mixed results. The first is [EDoc], a study on electronic documents using a *loLevel* Linear interface, a *embedded* Fisheye interface, and a *separate* Overview+Detail interface Hornbæk and Frokjær [2001], Hornbæk et al. [2003]. An essay task and a question-answering task were included. [EDoc] illustrates how task nature could affect the levels of data required, and how that difference could affect interface effectiveness. In the question-answering task, participants were slower, without being more accurate in their answers, if they were given an additional high-level view. Based on reading patterns, Hornbæk and Frokjær [2001] suggested that the slower reading times were due to the attention-grabbing, high-level view in the *separate* interface, which led participants to further explore the documents perhaps unnecessarily. In contrast, in the essay-writing task where participants were required to summarize the documents, having the extra high-level overview displaying data structure as section and subsection headers resulted in better quality essays without any time penalty when compared to the *loLevel* interface. In other words, when the task required single-level answers, as in the question-answering task, having an extra high-level display had a time cost; when the task required multi-level answers, as in the essay-writing task, the high-level display produced higher quality results.

The second study that showed mixed results is [BigOnSmall], which examined the effectiveness of three large-screen display techniques on small screens: *loLevel* Panning, *embedded* Fisheye, and *temporal* Two-level zoom Gutwin and Fedak [2004b]. Participants were asked to edit a PowerPoint figure in the *Editing* task and to detect device failures in the *Monitoring* task. Both of their multi-level interfaces provided performance benefits in these tasks, probably since the needed information was concentrated on the high-level displays. Indeed, Gutwin and Fedak [2004b] reported that "people were often able to carry out most of the task without zooming in at all" (p. 151). However, when the task required detailed information only available on the low-level displays, as in their *Navigation* task where participants were required to read text on a web page, their *temporal* Two-level zoom interface did not demonstrate any benefits over their *loLevel* Panning interface, probably because of the high costs of switching typically found in temporal interfaces. Their *embedded* Fisheye interface did demonstrate benefits, however. Since we did not have enough information regarding how the *embedded* Fisheye interface was used by their study participants, we cannot further explain how their *embedded* interface supported the *Navigation* task.

5.3   SUMMARY OF CONSIDERATIONS IN CHOOSING BETWEEN A SINGLE OR A MULTI-LEVEL INTERFACE

In general, the decision is made based on the benefits in display flexibility with multi-level displays and the amount of interaction effort required to coordinate multiple display levels. We found that when added visual levels did not add task-relevant information, as in the case of using multiple visual levels to display single-level data, costs incurred in visual-level coordination were typically not justified.

CHAPTER 6

Decision 2: How to Create the High-Level Displays?

Once the designer decides on taking a multi-level approach, the next step in the process is to create high-level displays. Creating high-level displays in a multi-level interface is a non-trivial task, especially when the amount of data involved is large. The consideration here is to provide the correct amount of data required by the tasks in forms that are perceivable and trusted by the users.

Study results suggest a delicate balance between displaying enough visual information for high-level displays to be useful and showing irrelevant resolution or information that becomes distracting. In Chapter 6.1, we discuss the adverse effects of displaying more visual levels than supported by the data and required by the task. Chapter 6.2 discusses the related topic of displaying too much information on high-level displays.

Given the space constraints, designers usually need to find less space-intensive visual encodings for the data or reduce the number of data displayed on high-level displays. Chapter 6.3 discusses cases where the researchers have gone too far in their visual-encoding abstraction as their study participants could no longer use the visual information on high-level displays. Chapter 6.4 looks at the trade offs in using *a priori* automatic filtering to selectively show data on high-level displays. Given all these considerations, we round up the discussion in Chapter 6.5 by re-examining the roles of high-level displays to help ground design. Study results suggest a more limited set of roles than proposed in the literature. While we found that study results supported the use of high-level views in *separate* interfaces as navigational shortcuts to move within the data and to provide overall data structure, we failed to find support for the common beliefs of using high-level regions in *embedded* interfaces to aid orientation or to provide meaning for comparative interpretation of an individual data value.

6.1   HAVING TOO MANY VISUAL LEVELS MAY HINDER PERFORMANCE

In general, the number of visual levels supported by the interface should reflect the levels of organization in the data. Otherwise, users may need to pay the cost of coordinating between different visual levels without the benefit of rich information at each level. In cases where the extra levels were not useful for the task at hand, the irrelevant information could be distracting. These extra visual levels may at best be ignored, and at worst, may harm task performance.

Of the 22 studies reviewed, four looked at compound multi-level interfaces when an additional high-level view was added to an already multi-level interface (Table 6.1).

Table 6.1: Four papers had at least one compound multi-level interface, created by adding an additional high-level view to a multi-level interface.

Effect	Paper	High-level view added to...
No benefits	[FCScreen] Baudisch et al. [2002]	*temporal* zoom plus pan (z+p) display to create their overview plus detail (o+d) interface
	[RubNav] Nekrasovski et al. [2006]	*temporal* Pan&Zoom and their *embedded* Rubber Sheet Navigation interfaces
Adverse effects	[ZuiNav] Hornbæk et al. [2002]	*temporal* zoomable interface
Excluded	[FishMenu] Hornbæk and Hertzum [2007]	*embedded* fisheye menu

Since [FishMenu] did not include an interface that was only *embedded* without the high-level overview, we could not discern the effects of having an additional high-level view and thus excluded it from this discussion. For the other three studies, perhaps because the multi-level interfaces already displayed all the meaningful and task-relevant visual levels supported by the data, having the additional high-level view did not enhance participant performance and sometimes degraded it.

Two studies showed a lack of benefit in providing additional high-level views (Table 6.1). Participants in [FCScreen] obtained similar performance using the overview plus detail (o+d) interface and their zoom plus pan (z+p) interface Baudisch et al. [2002]. [FCScreen] reported that participants kept the *temporal* view zoomed to 100% magnification for tracing, thus effectively reduced the *temporal* component of the interface to a single-level display, and used the compound multi-level interface as a *separate* interface (high-level display plus the *temporal* display used as a *loLevel* display).

[RubNav] reported a study on large trees and visual comparison tasks Nekrasovski et al. [2006]. Their high-level view showed an overall tree view and provided task-relevant location cues. However, the information was not unique and necessary as the low-level view also provided a similar visual cue. As a result, the study failed to show performance benefits in having an extra high-level view in their interfaces even though participants reported that the high-level view reduced physical demand.

A map-navigation study reported in [ZuiNav] suggested performance was hindered when an interface provided irrelevant levels of resolutions Hornbæk et al. [2002]. One of their study interfaces was a *temporal* interface with an added high-level overview. [ZuiNav] reported that participants who actively used the high-level overview had higher performance time, possibly because of the mental and motor efforts required in integrating the low- and high-level windows. Such costs were not compensated by richer information displays as the *temporal* interface already contained all the task-relevant visual resolutions and may have reduced, or even eliminated, the need for a separate overview (p. 381).

In some cases, study results indirectly suggested adverse effects on performance when the interfaces provided irrelevant visual levels. One example is [VisMem] Plumlee and Ware [2006]. The task was to match three-dimensional object clusters. The *temporal* interface had many magnification levels that neither helped participants to locate candidate objects, nor were detailed enough for visual matching. Given that participants needed to memorize cluster objects between temporal view switching in the *temporal* interface, the extra zooming levels may have rendered the tasks harder. This extra cognitive load may explain the relatively small number of items participants could handle before the opponent *separate* interface supported better performance, when compared to results obtained in [TimeGr] Saraiya et al. [2005].

Similarly, in [FCScreen], a study that looked at static visual path-finding tasks and dynamic obstacle-avoidance tasks, two of the study interfaces (*temporal* and *separate*) seemed to have included more visual levels than their *embedded* interface Baudisch et al. [2002]. While the special setup in [FCScreen]'s *embedded* interface undoubtedly contributed to the superior participant performances, we did wonder if the extra visual levels may have distracted participants in the other two interface trials.

6.2 HAVING TOO MUCH INFORMATION ON HIGH-LEVEL DISPLAYS MAY HINDER PERFORMANCE

While it may be tempting to provide more rather than less information on high-level displays, study results suggest that the extra information may harm task performance. None of the 22 reviewed studies included item density on high-level displays as a factor. However, we obtained indirect evidence by comparing between multi-level interfaces that display different amounts of visual information in their high-level displays and by comparing between low- and high-level displays for visual search tasks that only require information from high-level displays.

6.2.1 COMPARE BETWEEN HIGH-LEVEL DISPLAYS WITH DIFFERENT AMOUNTS OF VISUAL INFORMATION

As shown in Table 6.2, 17 of the 22 studies included at least two multi-level interfaces. Of the 17 studies, 13 displayed similar amounts of information on the high-level displays and could not be used to understand effects of showing task-irrelevant information on high-level displays. We excluded [eDoc] since their high-level displays showed different kinds, rather than different amounts, of information Hornbæk and Frokjær [2001], Hornbæk et al. [2003]. We also excluded [SPACETREE] since it was unclear from the paper the number of items initially shown in their *embedded* SpaceTree interface Plaisant et al. [2002].

Table 6.2: Seventeen papers included at least two multi-level interfaces. We compared the amounts of information displayed on the different high-level displays to understand effects of showing task-irrelevant information.

Amount of Low-level Info	Papers
Similar (excluded)	[FCScreen] Baudisch et al. [2002]
	[Fishnet] Baudisch et al. [2004]
	[DateLens] Bederson et al. [2004]
	[ElideSrc] Cockburn and Smith [2003]
	[FishSteer] Gutwin and Skopik [2003]
	[BigOnSmall] Gutwin and Fedak [2004b]
	[ZuiNav] Hornbæk et al. [2002]
	[LineGr] Lam et al. [2007]
	[RubNav] Nekrasovski et al. [2006]
	[VisMem] Plumlee and Ware [2006]
	[FishRadar] Schafer and Bowman [2003]
	[FishNav] Schaffer et al. [1996]
	[SpaceFill] Shi et al. [2005]
Different	[FishMenu] Hornbæk and Hertzum [2007]
	[InfoScent] Pirolli et al. [2003]
Excluded	[eDoc] Hornbæk and Frokjær [2001];
	Hornbæk et al. [2003]
	[SpaceTree] Plaisant et al. [2002]

Our discussion here therefore focuses on the two studies that displayed similar kinds of information, but at different amounts, on their high-level displays:

1. [INFOSCENT] compared the *separate* file browser with the *embedded* hyperbolic tree browser Pirolli et al. [2003]. While [INFOSCENT] did not explicitly compare display capacities of the two high-level displays, we estimated the amount of data displayed based on paper figures. The high-level view of the *separate* file browser displayed about 30 items. In contrast, the capacity of the high-level region of the *embedded* hyperbolic tree browser was at least two orders of magnitude larger.

2. [FISHMENU] compared the *temporal* cascading menu to two *embedded* menu designs based on the Fisheye menu Bederson [2000]. While the highest level of their *temporal* cascading

menu only showed a list of alphabets, their *embedded* fisheye menus showed all menu items in font sizes based on relative distances from the focus Hornbæk and Hertzum [2007].

In both cases, researchers advised against putting too much visual information on the display. Pirolli et al. [2003] argued against the assumption of " 'squeezing' more information into the display 'squeezes' more information into the mind" (p. 51) since visual attention and visual search interact in complex ways Pirolli et al. [2003]. In fact, [INFOSCENT] showed detrimental effects of display crowding. Pirolli et al. [2003] quantified information relevance as information scent. For their tree data set, they developed an Accuracy of Scent score, which was related to "(a) the ability of users to discriminate the information scent associated with different subtrees to explore and (b) the correctness of those choices with respect to the task" (p. 31). [INFOSCENT] found that their *embedded* hyperbolic tree browser interface led to slower performance times when compared to their *temporal* file browser under low information scent, possibly because their *embedded* interface displayed irrelevant information that was distracting.

Hornbæk and Hertzum [2007] came to a similar conclusion in [FISHMENU] on displaying menus with large numbers of items: "designers of fisheye and focus + context interfaces should consider giving up the widespread idea that the context region must show the entire information space" (p. 28) Hornbæk and Hertzum [2007]. We excluded their *temporal* cascading menu results in this discussion since their *separate* and their *embedded* interfaces had severe usability problems, and were therefore not comparable to the *temporal* results. We therefore focused on the two *embedded* interfaces and compared between them instead. Their Multifocus menu displayed larger numbers of readable menu items than the Fisheye menu, but had lower coverage of the data set. Eye-tracking results indicated that participants made more use of context and transition regions in the Multifocus menu than with the Fisheye menu. Hornbæk and Hertzum [2007] thus suggested dispensing with the unreadable, and therefore inaccessible, transition regions in the Fisheye menu (p. 26).

6.2.2   COMPARE BETWEEN HIGH- AND LOW-LEVEL DISPLAYS

When task answers are apparent from the high-level display, extra information in the *loLevel* display is therefore irrelevant. As shown in Table 6.3, nine of the 22 reviewed studies included a *loLevel* and at least one multi-level interface. Seven of them included tasks that could be answered using the high-level displays alone. We therefore attempted to understand effects of displaying unnecessary information by comparing participant performances of their multi-level interfaces, where participants were likely to have consulted mainly the high-level displays and their *loLevel* interfaces, where participants needed to sieve through irrelevant information to locate task answers. Except in the case of [LINEGR] and [SNAP] where a *hiLevel* interface was also studied, our findings were speculations as we could not be certain that participants focused on the high-level displays in the multi-level interfaces.

[FISHNET] studied information searches on web documents Baudisch et al. [2004]. In their *Outdated* task, participants were required to check if the web documents contained all four semantically highlighted keywords. In other words, the detailed readable content of the web documents

Table 6.3: Nine papers included a *loLevel* and at least one multi-level interface, classified by the locations from which participants could find answers to the tasks.

Task Answer Location	Papers
High level	[FISHNET] Baudisch et al. [2004]
	[ELIDESRC] Cockburn and Smith [2003]
	[BIGONSMALL] Gutwin and Fedak [2004b]
	[SUMTHUM] Lam and Baudisch [2005]
	[LINEGR] Lam et al. [2007]
	[SNAP] North and Shneiderman [2000]
	[TIMEGR] Saraiya et al. [2005]
Both high level and low level	[EDOC] Hornbæk and Frokjær [2001]; Hornbæk et al. [2003]
	[FISHSRC] Jakobsen and Hornbæk [2006]

displayed in their *loLevel* Linear interface was irrelevant to the *Outdated* task. Since their *separate* overview and their *embedded* Fishnet interfaces concentrated these task-relevant semantic highlights in their high-level displays, the two multi-level interfaces outperformed their *loLevel* interface for this task.

[ELIDESRC] studied program code navigation using the elision technique Cockburn and Smith [2003]. Their *Signature retrieval* task required participants to find data types of arguments in specific methods. For that task, only method definitions were needed. Since their *embedded* Illegible elision interface was the most efficient in concentrating method definitions, the interface was found to better support this task than their *loLevel* Flat-text interface.

[BIGONSMALL] examined two tasks that did not require information from low-level displays: the *Editing* task where participants were asked to edit a PowerPoint figure, and the *Monitoring* tasks where participants were asked to detect device failure Gutwin and Fedak [2004b]. In both cases, their multi-level interfaces demonstrated performance benefits over their *loLevel* Panning interface.

[SUMTHUM] studied information searches on web pages Lam and Baudisch [2005]. Their PDA-sized *temporal* interfaces supported equal performance as their desktop counterparts, even though the *loLevel* interface had nine times more display space showing completely readable information. The researchers suggested that the extra information on the desktop display may have distracted participants and caused unnecessary searching and reading, which may have resulted in the lack of performance benefits despite having a larger display.

Similarly, [TIMEGR] reported that their high-level, or single attribute, display was most helpful to analyze graphs at a particular time point, as "multiple attributes can get cluttered due to the amount of information being visualized simultaneously" (p. 231) Saraiya et al. [2005].

The following two studies also included a *hiLevel* display, thus enabling direct comparisons between low- and high-level displays.

[LINEGR] reported a study on visual target search in a large line-graph collection Lam et al. [2007]. One of the tasks involved finding the highest point in the data. The *hiLevel* interface alone was adequate for the task, and not surprisingly, interfaces that included a high-level display were found to support better performance than their *loLevel* interface. Observations suggested that about half of the participants did not use the low-level display in the multi-level interfaces for this task.

[SNAP] looked at visual information search on multiple views North and Shneiderman [2000]. Interfaces that were equipped with a high-level view (i.e., their *hiLevel* and *separate* interfaces) were found to be superior to the *loLevel* interface for tasks that could be answered based on information on these high-level views alone.

In short, instead of using physical item density as a measurement of space-use efficiency, a perhaps more useful consideration is the density of useful information on the display, which is arguably task or even subtask specific.

6.3   DISPLAYING INFORMATION IS NOT SUFFICIENT; INFORMATION HAS TO BE PERCEIVABLE

The mere presence of information on the screen is not sufficient; the information needs to be perceivable to be usable. Text on high-level displays may need to be readable to be useful. As shown in Table 6.4, nine of the 22 studies reviewed looked at text data. Six studies included unreadable text in their interfaces, while two had only readable text. We excluded [DATELENS] as both of their interfaces, the *embedded* DateLens and the *temporal* Pocket PC Calendar, used symbols to replace text in case of inadequate display area Bederson et al. [2004].

Table 6.4: Nine papers looked at text data, classified by the readability of the included text.

Text Readability	Papers
Some unreadable	[FISHNET] Baudisch et al. [2004]
	[ELIDESRC] Cockburn and Smith [2003]
	[BIGONSMALL] Gutwin and Fedak [2004b]
	[FISHMENU] Hornbæk and Hertzum [2007]
	[FISHSRC] Jakobsen and Hornbæk [2006]
	[SUMTHUM] Lam and Baudisch [2005]
Only readable text	[EDOC] Hornbæk and Frokjær [2001]; Hornbæk et al. [2003]
	[SNAP] North and Shneiderman [2000]
Excluded	[DATELENS] Bederson et al. [2004]

Study results showed that unreadable text displayed on high-level displays was not an effective shortcut to low-level details, as single *loLevel* displays resulted in similar participant performance despite displaying the information in a larger screen area and thus, having a larger search space.

[FISHNET] looked at information searches on web documents Baudisch et al. [2004]. Both of their multi-level interfaces showed unreadable text except for a few keywords. When the task required reading text in the neighborhood of these readable keywords, as in their *Analysis* task, the multi-level interfaces failed to demonstrate performance benefits over the traditional *loLevel* browser.

One of the tasks in [BigOnSmall] involved selecting links from web pages Gutwin and Fedak [2004b]. Text in the high-level display was illegible. Of the two multi-level interfaces tested, only the *embedded* Fisheye interface demonstrated time benefits over their *loLevel* Panning interface for this task.

[FishMenu] examined displaying large numbers of menu items Hornbæk and Hertzum [2007]. Their *embedded* Fisheye menu displayed unreadable items at the extreme ends in the high-level regions. Eye-tracking results indicated that participants made very little use of high-level regions, thus suggesting their ineffectiveness (p. 26).

[FishSrc] looked at displaying program code using an *embedded* fisheye interface, which displayed unreadable text in the high-level regions Jakobsen and Hornbæk [2006]. The *embedded* interface showed a time cost over the *loLevel* interface in a task that involved counting conditional and loop statements, as participants spent more time in the *embedded* interface to find closing braces of a loop control structures that were unreadable in the high-level regions. The researchers thus suggested that interfaces should display readable text to allow direct use of the high-level view information (p. 385).

[SumThum] reported similar findings Lam and Baudisch [2005]. Their *temporal* Thumbnail interface had unreadable high-level text, but their *temporal* Summary Thumbnail contained only readable high-level text. They found that participants using the Thumbnail interface had 2.5 times more zooming events, and when zoomed in, horizontally scrolled almost 4 times more, suggesting the ineffectiveness of the unreadable high-level text.

Perhaps one exception is [ElideSrc], a study on using elision to display program code 2003. Cockburn and Smith [2003] looked at two *embedded* interfaces. The *embedded* Legible elision interface showed program code in small but readable fonts, while their Illegible elision *embedded* interface showed elided code in one-point greeked letters that were unreadable. Surprisingly, for tasks that required reading method contents, having legible text did not demonstrate performance time advantages. According to Cockburn and Smith [2003], participants had difficulties reading the text in the *embedded* Legible elision interface, to the point where eight out of ten of their participants decided to expand the elided text to read method contents. It is therefore unclear if texts showed in their Legible elision interface was functionally readable.

For graphical visual signals, two studies reported effects of showing insufficient details on the high-level display Hornbæk et al. [2002], Lam et al. [2007]. In [ZuiNav], a study on map navigation, the geographic map information provided by the high-level overviews may not have been sufficiently detailed for the study tasks, for example, to find a neighboring location given a starting point, to compare the location or size of two map objects, or to find the two largest map objects in a geographic boundary Hornbæk et al. [2002]. For the Washington-map trials, having an extra high-level overview had time and recall accuracy costs, suggesting the burden of "switching between the detail and the overview window required mental effort and time moving the mouse" (p. 382). Indeed, "tasks solved with active use of the overview were solved 20% slower than tasks where the overview window was not actively used" (p. 380), possibly due to the insufficient information on the high-level overview that led to the large number of transitions between the overview and the detail window. Despite 80% of participants indicating subjective preference for having the extra view, only 55% of participants actively used the high-level view.

[LINEGR] qualified perceptual requirements for their high-level displays as visual complexity and visual span Lam et al. [2007]. The study looked at displaying a large collection of line graphs for visual search and visual compare tasks, and found that in order for the high-level view to be usable, the signal had to be visually simple and limited to a small horizontal area. For example, in the task that required finding the highest peak in the data collection, the visual signals on high-level displays were simple narrow peaks and could easily be found. In contrast, three-peak signals in their *Shape* task were complex and were less discernable in the high-level views. As a result, participants resorted to the low-level views for these three-peak signals.

In short, designers need to provide enough details for visual objects on high-level displays to be usable. For text, the display objects should be functionally readable if the tasks require understanding text content. For graphical objects, the criteria are less clearly defined.

6.4   *A PRIORI* AUTOMATIC FILTERING MAY BE A DOUBLE-EDGED SWORD

Table 6.5: Three papers implemented *a priori* automatic filtering. pos = positive effects observed; neg = negative effects observed.

Papers	Filtering Effect(s)
[FISHNET] Baudisch et al. [2004]	pos
[FISHSRC] Jakobsen and Hornbæk [2006]	pos and neg
[EDOC] Hornbæk and Frokjær [2001]; Hornbæk et al. [2003]	neg

Designers often can only display a subset of the data on the high-level displays. One selection approach is based on degree-of-interest function using *a priori* knowledge of data relevance with respect to the focus datum Furnas [1986]. Jakobsen and Hornbæk [2006] further differentiated the distance term in the function into semantic and syntactic distances to implement an *embedded* interface for source code. As seen in Table 6.5, of the three studies that implemented *a priori* automatic filtering, two suggested that automatic filtering could enhance task performance as high-level displays concentrated useful information and reduced distractors. However, in two studies, some participants were confused by the selective filtering and became disoriented.

6.4.1   FILTERING TO REMOVE IRRELEVANT INFORMATION

Instead of seeing filtering as a workaround to the display-size challenge and as a liability, there is evidence to suggest that filtering in itself can enhance task performance. When filtering selects task-relevant information for high-level display, such intelligence avoids tedious manual searching and navigation in low-level views, and possibly also avoids distractions by irrelevant information.

[FISHNET] studied information searches on web pages Baudisch et al. [2004]. Their multi-level interfaces semantically highlighted and preserved readability of keywords relevant to the tasks. These keywords were concentrated in smaller display spaces by reducing font sizes of surrounding text. Such interfaces resulted in better participant performances as long as they still provided task-required layout information. For example, participants were faster when using either of their multi-level interfaces for the *Outdated* task, and when using their web-column preserving *embedded* interface for the *Product-choice* task.

[FISHSRC] studied displaying program source code using a fisheye interface Jakobsen and Hornbæk [2006]. Automatic and semantically selected readable context in their *embedded* interface avoided the need to manually search for function declarations in the entire source code. This advantage manifested in faster performance times in tasks where participants were required to search for information contained in the function declarations throughout the entire source code.

6.4.2   FILTERING MAY CAUSE DISORIENTATION AND DISTRUST

However, automatic filtering may be a double-edged sword, as filtering may result in disorientation and distrust of the automatic selection algorithm. The *embedded* interface in [EDOC] preserved readability only for the most important part of the document, with content importance determined by the interface *a priori* Hornbæk and Frokjær [2001], Hornbæk et al. [2003]. Participants expressed distrust, both in their satisfaction feedback where they rated the *embedded* interface as confusing, and in their comments indicating that they "did not like to depend on an algorithm to determine which parts of the documents should be readable" (p. 142).

This problem may be worse with semantic filtering, where object visibility depends on the semantic relatedness of the object to the focus datum, rather than the geometric distance between screen displays. Selection of displayable context based on syntactic distance between the data point and the focus is arguably easier to predict than semantic selection. Consequently, it may be easier for users to understand and trust filtering algorithms based on syntactic distance only. Also, since context information is updated when the focal point changes, it may be more confusing to navigate with semantic-context updates, as pointer navigation is conceptually geometric rather than semantic. In [FISHSRC], high-level regions replaced scrolling in the *embedded* interface and only displayed semantically-relevant source code based on focus Jakobsen and Hornbæk [2006]. Participants were confused about the semantic algorithm that caused program lines to be shown and highlighted in the context area (p. 385) Jakobsen and Hornbæk [2006].

Another problem of automatic filtering is that the selection may affect the amount of time users spent on different parts of the data. [eDoc] reported that participants spent approximately 30% less time on the initially collapsed sections displayed on their *embedded* interface than when displayed in full on the other interfaces Hornbæk and Frokjær [2001], Hornbæk et al. [2003]. In short, while *a priori* filtering may concentrate task-relevant information on high-level displays, selective filtering may incur user distrust and confusion, and may even affect how users explore the displayed data.

6.5   ROLES OF HIGH-LEVEL DISPLAYS MAY BE MORE LIMITED THAN PROPOSED IN LITERATURE

While low-level displays enable users to perform detail work, the roles of high-level displays are harder to verify. We therefore looked at four proposed uses of high-level displays based on the published literature. We found that study results support two proposed claims concerning *separate* interfaces: high-level views provide navigation shortcuts and show overall data structure. We were unable to find strong support for using high-level regions in *embedded* interfaces to aid orientation or to provide meaning for data comparison.

6.5.1   SUPPORTED: HIGH-LEVEL VIEWS IN *SEPARATE* INTERFACES PROVIDE NAVIGATION SHORTCUTS

Information shown in the high-level views can facilitate navigation by providing long-distance links, thus "decreasing the traversal diameter of the structure" in navigation Furnas [2006]. Coordination between the low- and the high-level views enable users to directly select targets on high-level displays for detail exploration. For example, [Snap] found that the high-level view of a list of geographic states acted as hyperlinks for the high-level detail census data North and Shneiderman [2000].

Another way that high-level views assist navigation is by providing a map of available paths Card et al. [1999]. An example is the high-level overview in the *separate* interface in [eDoc] that showed section and subsection headers Hornbæk and Frokjær [2001], Hornbæk et al. [2003]. For graphical displays, [FCScreen] found that participants used the high-level overview to navigate to targets and performed the detail work in the *loLevel* display Baudisch et al. [2002].

High-level views can also be useful for refinding. [eDoc] reported reading pattern analysis showed that participants "used the overview pane to directly jump back to previously visited targets" and "the overview pane supports helps reader [sic] memorize important document positions" (p. 145) and resulted in participant preference and satisfaction even though this apparent navigation advantage failed to materialize as time performance benefits Hornbæk and Frokjær [2001], Hornbæk et al. [2003].

6.5.2   SUPPORTED: HIGH-LEVEL VIEWS IN *SEPARATE* INTERFACES SHOW OVERALL DATA STRUCTURE

High-level views can provide information about data structures that may not be apparent when the data are viewed in detail. For example, [εDOC] found that document section and subsection headers shown in the high-level view of their *separate* interface "may indirectly have helped subjects to organize and recall text" (p. 144), and led to a higher quality essay without any time penalty Hornbæk and Frokjær [2001], Hornbæk et al. [2003].

6.5.3   OPEN: HIGH-LEVEL REGIONS IN *EMBEDDED* INTERFACES AID ORIENTATION

When the information space contains little or no information for which we can base our navigational decisions, the problem of "desert fog" occurs Jul and Furnas [1998]. Global context in *embedded* displays has been proposed to help users orient Nigay and Vernier [1998], perhaps by providing visual support for working memory as the display gives evidence of where to go next Card et al. [1999].

While we did not find evidence to study this role of *embedded* high-level regions, results from Hornbæk et al. [2002] may shed some lights on the topic.

Results from [ZuINav] suggested that visual cues in data could aid navigation Hornbæk et al. [2002]. In [ZuINav], the Washington map contained rich visual cues for navigation. Participants were faster in navigation tasks performed using their *temporal* interface with the Washington map without the high-level view, suggesting that the map contained visual objects that aided navigation. In contrast, participants using the Montana map made a smaller number of scale changes when the high-level display was present, suggesting that the map itself did not contain enough visual objects for effective navigation, and participants needed the guidance of the high-level overview.

If visual objects displayed in high-level regions of *embedded* interfaces act similarly to navigational cues in the Washington map, it would be likely that high-level regions can aid orientation.

6.5.4   OPEN: HIGH-LEVEL REGIONS IN *SEPARATE* INTERFACES PROVIDE DATA MEANING

It is believed that data value is only meaningful when interpreted in relation to surrounding entities, and "the surrounding entities at different scales of aggregation exert a semantic influence on any given item of interest" Furnas [2006]. Again, we did not find *embedded* results to study this role. However, [TimeGr], a study on displaying time-series data as nodes in a graph, may provide some understanding.

[TimeGr] included a *loLevel* interface that showed all 10 time points simultaneously and a *temporal* interface that showed one data point at a time Saraiya et al. [2005]. Even though participants made more errors overall when using the *loLevel* interface, thus suggesting having surrounding entities may be detrimental rather than helpful, a closer look at individual tasks showed mixed results.

We focused on tasks that involved all time points as they were more likely to involve comparative interpretations. [TIMEGR] reported the *temporal* interface supported faster task time in finding the topology trend of a larger graph and in searching for outlier time points. These two results suggested that despite having to identify trends or detect outliers, context provided in the *loLevel* interface was detrimental rather than beneficial, possibly due to visual clutter. On the other hand, participants achieved better performance results with the *loLevel* interface for the two tasks that involved finding outlier nodes and groups, and they did not exhibit any performance differences for tasks that involved finding time trends.

Given the mixed results from [TIMEGR], we were unable to offer insights into the role of high-level regions in providing data meaning for comparison.

6.6   SUMMARY OF CONSIDERATIONS IN HIGH-LEVEL DISPLAY CREATIONS

Creating high-level displays is the second step in our decision tree (Figure 1.1). The consideration is to provide enough visual information for the tasks in forms that are usable and trusted by the users. Providing enough information implies matching the number of visual levels in an interface with the number of levels in the displayed data as extra visual levels may hinder performance. Similarly, high-level displays should only show task-relevant information, as extra information may be distracting. Information displayed should be perceivable in order to be useful. For text, readability is an important consideration; for graphical objects, the definition is less clear. Oftentimes, there are more items in the data than can be accommodated on the output device. Even though *a priori* selection of display data is an attractive solution, study results have found that doing so could lead to user confusion and distrust.

CHAPTER 7

Decision 3: Simultaneous or Temporal Displays of the Multiple Visual Levels

The third decision in the process of creating a multi-level interface is on visual level arrangements. For the designer, it is a choice between showing the levels simultaneously or one at a time, as in zooming techniques.

A well-known problem with zooming is that when the user zooms in on a focus, all contextual information is lost. Loss of context can be a considerable usability obstacle, as users need to integrate all information over time, an activity that requires memory to keep track of the temporal sequence and their orientations within that sequence Furnas [2006], Herman et al. [2000]. To alleviate these problems, a set of techniques collectively called focus + context were developed. Indeed, Card et al. [1999] stated the first premise of focus + context visualization as that "the user needs both overview (context) and detail information (focus) simultaneously" (p. 307). Another problem of zooming is that it "'uses up' the temporal dimension—making it poor for giving a focus + context rendering of a dynamic, animated world" Furnas [2006].

Although this reasoning appears to be logical, empirical study results did not consistently support using simultaneous level displays: study results suggested that the *temporal* interface was surprisingly good for most tasks. We identified two situations where the simultaneous-level display provided performance benefits: when the answer to the problem involved information from **all** the available levels (Chapter 7.1) and when the different levels provided clues for the task (Chapter 7.2). Otherwise, temporal switching seemed adequate as discussed in Chapter 7.3. In short, the consideration appears to be a balance between interaction costs of the display and potential benefits, which hinge on the number of data levels required by the tasks.

7.1 TASKS WITH MULTI-LEVEL ANSWERS BENEFITED FROM SIMULTANEOUS DISPLAYS OF VISUAL LEVELS

In general, we found that the display of simultaneous levels was best suited for tasks that required multi-level answers. Twelve of the 22 studies included a *temporal* and at least one simultaneous-display interface for comparison (Table 7.1). We excluded three studies in this discussion:

Table 7.1: Twelve papers included a *temporal* and at least one simultaneous-display interface. MC = Multiple-level clues; SC = Single-level clues; SB = Single-level interface better supported tasks.

Papers	MC	SC	SB
Multi-level answers			
[DateLens] Bederson et al. [2004]			
[SpaceTree] Plaisant et al. [2002]			
[FishNav] Schaffer et al. [1996]			
Single-level answers			
[FCScreen] Baudisch et al. [2002]	x		
[DateLens] Bederson et al. [2004]		x	x
[BigOnSmall] Gutwin and Fedak [2004b]		x	
[RubNav] Nekrasovski et al. [2006]		x	x
[InfoScent] Pirolli et al. [2003]	x		
[SpaceTree] Plaisant et al. [2002]	x		
[VisMem] Plumlee and Ware [2006]	x		
[FishNav] Schaffer et al. [1996]		x	x
[SpaceFill] Shi et al. [2005]		x	
Excluded			
[ScatterPlot] Buring et al. [2006a]			
[ZuiNav] Hornbæk et al. [2002]			
[FishMenu] Hornbæk and Hertzum [2007]			

1. [ScatterPlot] looked at displaying scatterplots on small screens Buring et al. [2006a]. The study found no significant differences between the test interfaces, possibly due to implementation-dependent usability issues.

2. [ZuiNav] looked at the effects of an added high-level overview to a zoomable user interface for map navigation Hornbæk et al. [2002]. Their *separate* interface, the zoomable interface with an overview, was effectively used as just a *temporal* interface most of the time;

3. [FishMenu] looked at the usability of fisheye menus showing 100 and 292 items Hornbæk and Hertzum [2007]. Their simultaneous-level interfaces had various implementation-dependent usability issues.

Three of the nine included studies had at least one task that required multi-level answers, and all showed performance benefits in using their simultaneous-display interfaces for those tasks compared to their *temporal* interfaces.

The *embedded* DateLens interface in [DATELENS] was found to be more effective than the *temporal* Pocket PC interface for tasks that involved counting events within a 3-month time period in the calendar, for example, in counting scheduled events or appointment conflicts Bederson et al. [2004].

In [SPACETREE], the *embedded* SpaceTree interface trials were faster than the *temporal* Explorer interface on average and more accurate in a task that required listing all the ancestors of a node Plaisant et al. [2002].

In the re-routing task in [FISHNAV], participants were required to find an alternative route to connect two points in the network that were disconnected, and the route spanned all levels in the hierarchical network Schaffer et al. [1996]. The *embedded* interface supported faster task completion times and required only half the number of zooming actions when compared to the *temporal* interface. The advantage of the *embedded* interface could be its display of the ancestral nodes along with the children nodes at the lowest level of the hierarchy since all of which were needed to find an alternative route.

On the other hand, the *temporal* interface seemed to offer better support for tasks with single-level answers unless the clues required to reach the answers were also multi-level as discussed in the next section.

7.2   TASKS WITH MULTI-LEVEL INFORMATION CLUES BENEFITED FROM SIMULTANEOUS DISPLAY OF VISUAL LEVELS

For tasks with single-level answers, simultaneous-level display was still helpful if the clues to the tasks spanned multiple data levels. As shown in Table 7.1, four of the nine included studies had multi-level clues to single-level answers. All these studies demonstrated benefits in using simultaneous-level displays.

Multi-level interfaces in [FCSCREEN] supported equal or better performance than their *temporal* interface in the route-finding and connection-verification tasks Baudisch et al. [2002]. Even though the answer could be obtained in the low-level view alone, both tasks required global relative locations in high-level displays and detail information in low-level displays.

[INFOSCENT] looked at a similar phenomenon called information scent Pirolli et al. [2003]. Study results suggested that the *embedded* hyperbolic tree interface may support faster task time than the *temporal* file explorer interface at high-scent tasks. In their *embedded* hyperbolic interface, participants could see more of the hierarchical structure in a single view and traversed tree levels faster. Under high-scent conditions where ancestor nodes provided clues to task answers, this feature could be advantageous. In contrast, under low information scent conditions, participants examined more tree nodes when using the *embedded* than the *temporal* interface, resulting in slower task times.

[SPACETREE] reported that the *embedded* SpaceTree supported equal or better task times in the first-time tree node finding tasks than the *temporal* Explorer interface Plaisant et al. [2002]. Even though Plaisant et al. [2002] did not provide enough task instructions for us to judge if the task provided multiple-level clues, they did mention providing hints to participants that seemed to span multiple levels: "To avoid measuring users' knowledge about the nodes they were asked to find (e.g. kangaroos) we provided hints to users (e.g., kangaroos are mammals and marsupials) without giving them the entire path to follow (e.g., we didn't give out the well known step such as animals)." (p. 62).

The task in [VisMem] required matching complex clusters of three-dimensional objects Plumlee and Ware [2006]. Clues to the answers were present in both the high-level view, showing the location of the candidate targets, and in the low-level view, showing the details required in visual matching. Their *separate* interface was found to better support the task when the total number of objects per cluster was above five items, in which case participants could no longer hold all the clues in their short-term memory when using the *temporal* interface.

7.3   TASKS WITH SINGLE-LEVEL INFORMATION CLUES MAY BE BETTER WITH TEMPORAL SWITCHING

Taking the previous two considerations together, we concluded that tasks with single-level answers and single-level clues would not benefit from simultaneous visual level displays. Indeed, study results seemed to support this observation even for the tasks that required object comparisons: As long as participants could keep task-required information in their short-term memory, *temporal* interfaces seemed adequate, and at times, even resulted in better participant performances and feedback.

As shown in Table 7.1, five of the nine included studies had at least one task that required single-level answers and provided single-level clues. Three of them supported this conclusion ([DateLens], [RubNav], and [FishNav]), while two do not ([BigOnSmall] and [SpaceFill]).

[DateLens] showed that the *temporal* Pocket PC was more appropriate for simple calendar tasks that involved checking start dates of pre-scheduled activities and tasks that spanned short-time periods Bederson et al. [2004].

[RubNav] studied a task that compared topological distances between coloured nodes in a large tree Nekrasovski et al. [2006]. Their results showed that their *temporal* interface outperformed their *embedded* interface even though the task required comparison between objects. Indeed, their *temporal* interface was rated by participants as being less mentally demanding and easier to navigate. In [FishNav], even though the *embedded* interface supported faster task times than *temporal* in rerouting within a hierarchical network, participants did not seem to need simultaneous-level display to locate broken links at the lowest network level, as indicated by the lack of performance differences between the *temporal* and the *embedded* interface trials for this link-location task Schaffer et al. [1996].

Two possible exceptions to this conclusion are found in [BigOnSmall] and [SpaceFill].

The *Navigation* task in [BigOnSmall] required participants to select and click on links on web pages Gutwin and Fedak [2004b]. Since text was illegible in the high-level displays, it is arguable that the high-level display did not provide enough information clues to the participants and thus, the interface only provided single-level clues at the low visual level displays. Study results showed benefits for using the *embedded* Fisheye interface over the *temporal* Two-level zoom interface for the *Navigation* task despite the single-level clue. Implementation details may explain study findings: zoom-level switching was performed with a key combination in the *temporal* interface, while a mouse click was presumably used in the *embedded* case, thus making the *temporal* display more difficult to use.

The second possible exception is the study in [SPACEFILL], a study on browsing using hierarchical space-filling visualizations. [SPACEFILL] reported that their *embedded* interface supported faster task times than the *temporal* interface Shi et al. [2005]. In this case, there may be a speed-accuracy tradeoff: Shi et al. [2005] observed that in some cases, their participants ignored potential targets that occupied a small amount of space and missed the small targets in less than 3.75% of the trials. Even though [SPACEFILL] did not report task error rates, the paper reported that this phenomenon may have a more severe and adverse impact on their *embedded* than on their *temporal* interface trials. Also, there were participants who gave up when using the *embedded* interface, but they only timed-out in the *temporal* trials. It was therefore unclear to us whether the *embedded* interface was truly superior to the *temporal* interface for these browsing tasks.

In short, simultaneous-level display is appropriate for multi-level answers or single-level answers found by multi-level clues. Otherwise, the *temporal* interface seemed adequate.

7.4 CONSIDERATIONS IN CHOOSING BETWEEN TEMPORAL SWITCHING OR SIMULTANEOUS DISPLAY OF THE VISUAL LEVELS

In general, simultaneous level display, as in *embedded* or *separate* interfaces, requires more complex interactions, while *temporal* interfaces can be taxing on the user's memory. Study results suggested that temporal switching was more suitable for tasks that did not involve multi-level answers, or did not provide multi-level clues to single-level answers. The decision is therefore based on the number of task relevant data levels.

CHAPTER 8

Decision 4: How to Spatially Arrange the Visual Levels, Embedded or Separate?

The last step in our decision tree is to decide between the two spatial arrangements of simultaneous-level display: the interface can embed the different levels within the same window or show them as separate views. Proponents of the embed approach argued that the different levels should be integrated into a single dynamic display, much as in human vision Card et al. [1999], Furnas [2006]. View integration is believed to facilitate visual search as it provides an overview of the whole display, which "gives cues (including overall structure) that improve the probability of searching the right part of the space" (p. 22) Pirolli et al. [2003]. Integrated views of data are argued to "support and improve perception and evaluation of complex situations by not forcing the analyst to perceptually and cognitively integrate multiple separate elements" (p. 83) Thomas and Cook [2005]. Also, it is believed that when information is broken into two displays (e.g., legends for a graph, or overview + detail), visual search and working memory consequences degrade performance as users need to look back and forth between the two displays Card et al. [1999], Pirolli et al. [2003]. On the other hand, spatial embedding frequently involves distortion, an issue discussed in Chapter 8.1.

Table 8.1: Eight papers included both *embedded* and *separate* interfaces, classified by participant performances. ex = excluded.

Embedded vs. *separate*	Papers
Unable to compare (ex)	[FISHNET] Baudisch et al. [2004]
	[EDOC] Hornbæk and Frokjær [2001]; Hornbæk et al. [2003]
	[RUBNAV] Nekrasovski et al. [2006]
No difference	[FISHMENU] Hornbæk and Hertzum [2007]
	[LINEGR] Lam et al. [2007]
	[FISHRADAR] Schafer and Bowman [2003]
embedded better	[FCSCREEN] Baudisch et al. [2002]
	[FISHSTEER] Gutwin and Fedak [2004a]

The choice between these two spatial arrangements is unclear based on empirical study results. Oftentimes, perceived functions of the two interfaces biased study data and task selections. For example, studies tended to use trees or graphs for node finding to study *embedded* interfaces (e.g., Plaisant et al. [2002]; Pirolli et al. [2003]; and Shi et al. [2005]) and spatial navigation for *separate* displays (e.g., North and Shneiderman [2000] and Plumlee and Ware [2006]). As a result, the issue of spatial arrangement was frequently confounded in our reviewed studies.

As shown in Table 8.1, 8 of the 22 studies included both *embedded* and *separate* interfaces. We found it difficult to directly compare between the two simultaneous displays in three of the studies ([Fishnet], [eDoc], and [RubNav]). These studies were thus excluded in our discussion.

[Fishnet] and [eDoc] were excluded due to intentional implementation differences based on common perceived use of the two spatial arrangements: high-level view in the *separate* interface to display data overview, and high-level regions in the *embedded* interface to show background and supporting information. [Fishnet], a study on web document search, included an *embedded* interface that was designed to favour row discrimination, while their *separate* interface favoured for column discrimination Baudisch et al. [2004]. This design choice thus added another factor that influenced study results and we therefore excluded [Fishnet] in this discussion.

Instead of differing layouts, [eDoc] showed different kinds of information in their two multi-level interfaces Hornbæk and Frokjær [2001], Hornbæk et al. [2003]. The high-level view of their *separate* interface showed document section and subsection headers and was optimal for displaying overall structure in text documents and for encouraging detail explorations. In contrast, their *embedded* interface showed *a priori* determined text significant to the focal area, which promoted rapid document reading at the cost of accuracy. Due to this intentional difference in interface use, we excluded [eDoc] in this discussion.

The last study in the incomparable group did not intend to study spatial arrangement despite including both *separate* and *embedded* interfaces. [RubNav] studied large tree displays Nekrasovski et al. [2006]. The goal of their *separate* interface was to investigate the use of an extra high-level view. Consequently, neither of their *separate* interfaces (*temporal* with overview and *embedded* with overview) could be directly compared with their *embedded* interface to discern effects of spatial arrangements, and [RubNav] was thus excluded.

In the five cases where direct comparison was possible, three studies did not find performance differences between the two simultaneous interfaces ([FishMenu], [LineGr], and [FishRadar]). The two exceptions were [FishSteer] and [FCScreen].

Even though [FishSteer] showed significant differences between their *separate* and *embedded* interfaces in steering tasks, we believe their results may be confounded by the relatively complex interactions required in their *separate* interfaces Gutwin and Fedak [2004a]. [FishSteer] included three *embedded* fisheye displays and two *separate* displays. In a series of two-dimensional steering tasks where participants were required to move a pointer along a defined path, [FishSteer] found that the *embedded* interfaces supported better time and accuracy performances over the *separate* interface at all display magnifications. Gutwin and Fedak [2004a] thus concluded that "the fact that fisheyes show[ed] the entire steering task in one window clearly benefited performance" (p. 207).

However, we believe a number of factors were involved in addition to the different level spatial arrangements. The first factor was differing effective steering path widths and lengths between interfaces. Of the five study interfaces, only one of the *separate* interfaces, the Panning-view, had an increased travel length at higher magnifications. All other interfaces had constant control/display ratios over all magnifications. As for the Radar-view *separate* interface, participants interacted with the high-level miniature view instead of the magnified high-level view, thus the actual steering path width was effectively constant over all magnifications.

We also found that interaction complexity differed greatly among the five interfaces. The *separate* Panning-view interface in [FISHSTEER] had more complex panning interactions than the other interfaces, especially at higher levels of magnification of the steering path. The *separate* Panning-view interface required two mouse actions, mouse drag for panning and mouse move for steering, while the *separate* Radar-view interface required only mouse-drag on the miniature high-level view. In contrast, the *embedded* interfaces required only a single mouse action to shift the focal point and magnify the underlying path. This type of interaction, however, has the disadvantage of a magnification-motion effect where objects in the magnifier appear to move in the opposite direction to the motion of the lens, so it is easy to overshoot the motion and slip off the side of the lens. We considered this motion effect as a third factor in the study.

Given the complex interplay of at least three factors that seemed to be implementation specific, we failed to extract general conclusions on visual level spatial arrangement based on [FISHSTEER].

[FCScREEN] looked at three tasks that required information from all levels: a static route-finding task, a static connection-verification task, and a dynamic obstacle-avoidance task Baudisch et al. [2002]. Study results indicated that the *embedded* interface better supported all of the tasks and was preferred by participants. Their unique *embedded* interface implementation avoided many of the usability pitfalls in embedding high-level regions into high-level displays, which may explain its superior participant performance: first, the location for the low-level region was fixed, thus potentially avoiding disorientation of a mobile focus with respect to the context area and the associated complex interactions, and second, distortion was not used in the system. Instead, Baudisch et al. [2002] used different hardware display resolutions for the two different levels. In contrast, their *separate* interface seemed more interactively complicated than the usual implementation, requiring panning in both low- and high-level views and zooming in the high-level view. Nonetheless, we believe their study demonstrated an effective use of their *embedded* interface over their *separate* interface.

We conclude that there is not sufficient evidence to derive design guidelines in choosing between the two simultaneous displays, as it is difficult to draw conclusions based only on [FCScREEN].

8.1   THE ISSUE OF DISTORTION

One of the potential costs in embedding multiple visual levels within the same window is distortion. Based on fisheye views Furnas [1986] and on studies of attention, Card et al. [1999] justified distortion since "the user's interest in detail seems to fall away from the object of attention in a systematic way and that display space might be proportioned to user attention". Also, Card et al. [1999] reasoned that "it may be possible to create better cost structures of induced detail in combination with the information in focus, dynamically varying the detail in parts of the display as the user's attention changes [...] Focus and context visualization techniques are 'attention-warped' displays, meaning that they attempt to use more of the display resource to correspond to interest of the user's attention" (p. 307).

Even though distortion is believed to be justified, it is still useful to examine the costs. The first problem is that distortion may not be noticed by users and be misinterpreted Zanella et al. [2002], especially when the layout is not familiar to the user or is sparse Carpendale et al. [1997]. Even when users recognize the distortion, distance and angle estimations may be more difficult and inaccurate when the space is distorted Carpendale et al. [1997], except perhaps in constrained cases such as bifocal or modified fisheye distortions Mountjoy [2001]. Also, users may have difficulties understanding the distorted image to associate the components before and after the transformation Carpendale et al. [1997], or in identifying link orientation in the hyperbolic browser Lamping et al. [1995].

To our knowledge, only four sets of published studies measured effects of distortion directly and systematically. Lau et al. [2004] found that a nonlinear polar fisheye transformation had a significant time cost in visual search, with performance slowed by a factor of almost three under large distortions. In terms of visual memory costs, our own work found that image recognition took longer and was less accurate at high fisheye transformation levels Lam et al. [2006]. Skopik and Gutwin [2003] investigated how people remember object locations in connected graph distorted at various levels using a Sarkar-Brown fisheye lens. Their studies found that people used different strategies at different levels of distortions, and landmarks became more important at higher distortions. In another sets of studies, the same researchers reported a time penalty without compromising accuracy on refinding nodes in a highly-linked graph when the graph was transformed by a polar fisheye transformation Skopik and Gutwin [2005].

It was difficult to tease out the effects of distortion based on the 22 papers we reviewed here since none of the studies specifically looked at distortion as a factor. We could therefore only rely on observations reported in the papers to obtain insights. As shown in Table 8.2, 16 studies included an *embedded* interface, and 14 implemented distortion. The two exceptions were [FCSCREEN] and [LINEGR]. [FCSCREEN] took a hardware approach and implemented their *embedded* interface with two different pixel resolutions Baudisch et al. [2002] and [LINEGR] used two distinct visual encodings to represent the same data in two levels Lam et al. [2007].

Interestingly, not all 14 studies reported usability or performance problems with visual distortion. In fact, nine studies reported performance benefits in using their distortable interfaces. We excluded [FISHSTEER] in this analysis as we could not tease out the effects of distortion based on

Table 8.2: Sixteen papers included at least one *embedded* interface. pos = Performance benefits demonstrated; neg = Problems reported; ex = excluded from review.

Papers	Distortion			Effects	
	None	Text	Grid	pos	neg
[FCSCREEN] Baudisch et al. [2002]	x				
[FISHNET] Baudisch et al. [2004]		x		x	
[DATELENS] Bederson et al. [2004]			x	x	
[ELIDESRC] Cockburn and Smith [2003]		x		x	
[FISHSTEER] Gutwin and Skopik [2003]**(ex)**				x	
[BIGONSMALL] Gutwin and Fedak [2004a]			x	x	
[EDOC] Hornbæk and Frokjær [2001]		x		x	
[FISHMENU] Hornbæk and Hertzum [2007]**(ex)**		x			x
[FISHSRC] Jakobsen and Hornbæk [2006]		x		x	
[LINEGR] Lam et al. [2007]	x				
[RUBNAV] Nekrasovski et al. [2006]					x
[INFOSCENT] Pirolli et al. [2003]					x
[SPACETREE] Plaisant et al. [2002]					x
[FISHRADAR] Schafer and Bowman [2003]					x
[FISHNAV] Schaffer et al. [1996]			x	x	
[SPACEFILL] Shi et al. [2005]			x	x	

study results due to the large number of factors involved in the study, as discussed earlier in this chapter. The remaining eight studies that demonstrated positive effects of distortion involved either text or grid-based distortions, suggesting that constrained and predictable distortions were well tolerated.

Five studies reported problems attributed to distortion, and all involved comparatively more drastic and elastic distortion techniques than text or grid-based distortions. We also excluded [FISHMENU] in our analysis since, even though the researchers reported usability problems with their various *embedded* and *separate* interfaces, it is unclear how distortion contributed to these problems. We therefore focus our discussion on the remaining four studies to further understand distortion costs: [RUBNAV], [SPACETREE], [INFOSCENT], and [FISHRADAR].

[RUBNAV] implemented an *embedded* Rubber-Sheet Navigation interface that allowed users to stretch or squish rectilinear focus areas as though the data set was laid out on a rubber sheet with its borders nailed down Sarkar et al. [2003]. Nekrasovski et al. [2006] attributed the relatively poor performance of their *embedded* interface to the disorienting effects of distortion (p. 18) Nekrasovski et al. [2006].

[SPACETREE] found that their participants took longer to refind previously-visited nodes in a tree using the *embedded* hyperbolic and SpaceTree interfaces than with the traditional *temporal* Microsoft Explorer file browser Plaisant et al. [2002] . Among the two *embedded* interfaces, participants demonstrated better performance with SpaceTree than with the hyperbolic tree browser, which involved drastic distortions. This result was predicted by the researchers as in SpaceTree, "the layout remains more consistent, [thus] allowing users to remember where the nodes they had already clicked on were going to appear, while in the hyperbolic browser, a node could appear anywhere, depending on the location of the focus point" (p. 62).

[INFOSCENT] also compared between a *temporal* file browser and the *embedded* hyperbolic tree browser Pirolli et al. [2003]. The researchers found that the hyperbolic tree browser supported better performance only for tasks with high information scent. Even though Pirolli et al. [2003] did not explicitly report problems related to distortion, they suggested providing landmarks to aid navigation in the *embedded* hyperbolic tree browser, thus indicating potential interaction costs in hyperbolic distortions.

[FISHRADAR] studied a radar fisheye view on maps as their *embedded* interface Schafer and Bowman [2003]. Their study reported both positive and negative effects of distortion. On the positive side, if noticed, the distortion enhanced awareness of the viewport in a collaborative traffic and sign positioning task using a map. However, users may not have noticed the distortion when it was caused by the actions of collaborators rather than their own direct actions.

In short, while we believe interfaces that implement distortions were generally more difficult to use, constrained and predictable distortions were found to be better tolerated and may tip the tradeoff between showing more information simultaneously on the display and the risk of causing disorientation and confusion.

8.2   CONSIDERATIONS IN SPATIALLY ARRANGING THE VARIOUS VISUAL LEVELS

In summary, there are trade offs in using either of the two simultaneous displays, *embedded* and *separate*. *Embedded* interfaces tend to implement distortion, which may be difficult for participants to understand and may involve difficult interactions, especially for the more drastic distortions such as the hyperbolic tree. For *separate* interfaces, view coordination has been found to be difficult.

CHAPTER 9

Limitations of Study

While we attempted to provide a comprehensive and objective systematic review in the use and design of multi-level interfaces, we were necessarily limited by our method. In this chapter, we discuss four major limitations of our study, including reviewer bias, misinterpretations, limited analysis scope, and qualitative recommendations.

9.1 REVIEWER BIAS

Our qualitative and bottom-up approach may suffer from reviewer bias in our study inclusion and in our emphasis, may put on various study results. In terms of study inclusion, we were further limited by our own resources, both in time and in knowledge.

To ensure objectivity, or at least to convey to our readers the basis of our claims, we listed the studies we considered in each of the design considerations. Given that we collected only 22 papers, we believe that explaining each set of study results qualitatively, instead of attempting statistical analysis, provides a more encompassing snapshot of our collective knowledge on multi-level interface use. While we did count the number of studies that produced statistically significant results towards each design consideration, we did not take the vote-counting approach in systematic reviews, as we did not base our findings on these numbers. Instead, we considered each study publication individually to identify evidence that might either support or refute our findings, taking into consideration possible explanations for study results beyond simply achieving statistical significance. In fact, we put more emphasis in the researchers' insights reported in their publications than on statistical results.

9.2 MISINTERPRETATIONS

We based our synthesis entirely on the reviewed publications to provide an evidence-based synthesis. In many cases, the goals of these reports were to directly compare interfaces as a whole, especially when one or more of the interfaces were novel. Given our goal to understand interface use, we often had to read the publications from a different perspective, and consequently, we may have misread or incorrectly inferred information from these publications.

9.3 LIMITED ANALYSIS SCOPE

Another consequence in basing our discussion entirely on our reviewed papers is limited scope: our discussions were limited by the factors reported in our reviewed publications, and our synthesis is necessarily incomplete. For example, while it may be more fruitful to look at how design principles behind study interfaces affected participants' objective performances, not all reviewed studies explicitly stated their design choices, either in their technique sections or in their evaluations. As a result, we were unable to meaningfully analyze design principles as implementation details are vital to the outcome of designs.

For the same reasons, our analysis at times may seem too high level and superficial as we oftentimes were unable to perform deeper analysis on the study results. Our difficulty lies partly in the diverse and disparate nature of the studies, and partly due to inconsistent and inadequate study reporting Lam and Munzner [2008]. As a result, we did not feel that we had enough data to relate our results to theoretical frameworks such as the Information Foraging Theory Pirolli [2007] except in passing.

9.4   QUALITATIVE RECOMMENDATIONS

We could only offer qualitative rather than quantitative design guidelines as results from our reviewed studies oftentimes did not systemically quantify design metrics. For example, Chapter 6.2 states "having too much information on high-level displays may hinder performance". While we provided examples from our reviewed studies to illustrate our point, we were unable to provide a quantitative metric for "too much information". The situation is similar in our discussion of distortion in Chapter 8.1 where we speculated on the types of distortion that would be best tolerated by participants when using *embedded* interfaces, but were unable to provide quantitative values on acceptable ranges of distortion.

CHAPTER 10

Design Recommendations

In this paper, we examine how the choice of interface elements (such as the number and organization of visual levels) hinges on the interface factors of data and tasks. In this chapter, we summarize our findings as three recommendations to designers in creating multi-level interfaces.

10.1 PROVIDE THE SAME NUMBER OF VISUAL LEVELS AS THE LEVELS OF ORGANIZATION IN THE DATA

Furnas [2006] argued for the need to provide more than two visual levels in his 2006 paper:

> By presenting only two levels, focus and context, these differ from the richer range of trading off one against the other represented in the canonical FE-DOI. This difference must ultimately prove problematic for truly large worlds where there is important structure at many scales. There the user will need more than one layer of context.

In the same paper, he also argued that the levels of resolutions can be determined based on the scale bandwidth of the presentation technology and scale range of the information world (p. 1003) Furnas [2006].

Looking at the question from a different angle, study results suggested that the effectiveness in providing multiple levels, especially simultaneous display of different levels, was contingent upon the number of organization levels in the data and the information needs of the task. In fact, we found that having extra levels may actually impede task performance, especially in *temporal* interfaces where users coordinate between the different levels using short-term memory. We believe that interfaces should therefore provide one visual level per data level.

10.2 PROVIDE RELEVANT, SUFFICIENT, AND NECESSARY INFORMATION IN HIGH-LEVEL DISPLAYS TO SUPPORT CONTEXT USE

While low visual levels should support detail work demanded by tasks at hand, study results suggested that high-level views in *separate* interfaces were used in two ways: in navigation where they provided short-cuts to jump to different parts of the data; and in mental data organization if they displayed overall data structure. To be effective, designers need to include only sufficient, relevant, and necessary information in the high-level views. This finding is in accordance with Norman [1993]'s Appropriateness Principle, where he stated that the visual representation should provide neither more or less information that is needed for the task at hand since extra information displayed may be distracting and render the task more difficult Norman [1993]. In the case of multi-level interfaces, displaying an inappropriate amount of information may tip the balance as the value of the display may not be sufficient to overcome the costs of having the extra visual resolutions. The amount of detail for each visual object required for high-level views is likely to be more than previously assumed in our community, judging from the number of ineffective high-level views created for the reviewed studies. For text documents, readability may be a requirement, as suggested in Jakobsen and Hornbæk [2006]: the design should "saturate the context area with readable information" in building interfaces to display program source code (p. 386), and in Hornbæk and Hertzum [2007]: "making the context region of the [fisheye menu] interfaces more informative by including more readable or otherwise useful information" (p. 28). For graphical displays, studies on visual search (e.g., Tullis [1985]) and Lam et al. [2007] provided guidelines, for example, visual signals should be simple and of narrow visual spans to be accessible, but the criteria still remain unclear.

10.3   SIMULTANEOUS DISPLAY OF VISUAL LEVELS FOR MULTI-LEVEL ANSWERS OR CLUES

Selecting the correct visualization technique to display data is important, due to the inherent trade offs in the *temporal*, *separate*, and *embedded* techniques. While most *temporal* implementations offer familiar panning and zooming interactions, these interfaces require users to keep information in their short-term memories. Simultaneous-level displays, on the other hand, often require more complex and unfamiliar interactions such as view coordination. Based on study results, we concluded that if the task or subtask needs information from multiple visual levels, either as part of the answer to the task or as clues leading to the answer, the interface should show these visual levels simultaneously. Otherwise, the *temporal* technique should be more suitable due to its simpler interface and more familiar interactions.

CHAPTER 11

Discussion and Future Work

We conducted a factor-based analysis of 22 existing multi-level interface studies to extract a detailed set of design guidelines that take into account the interaction of the interface factors of visual elements, data, and tasks. We cast our findings into a four-point decision tree for designers: (1) When are multiple visual levels useful? (2) How to create high-level displays? (3) Should the visual levels be displayed simultaneously? (4) Should the visual levels be embedded, or separated? In each of these steps, we examined how design decisions are affected by the interacting factors and demonstrated that design decisions cannot be based solely on study results at the monolithic interface level. One such example is the need for multiple visual levels in interfaces, which is generally considered to be beneficial in our community. However, empirical evidence suggests that when the task only requires data from a single organization level, the costs of managing multiple visual levels is not justified. Similar arguments apply in the question of the number of visual levels included in a visualization.

To our surprise, we are unable to suggest guidelines in displaying multiple levels simultaneously, either as *embedded* or as *separate* displays, due to the difficulties in obtaining direct interface comparisons based on our set of reviewed studies. While distortion in general involves more difficult interactions, constrained and predictable distortions such as text and grid-based distortions in general resulted in better participant performances than the more drastic distortions such as the hyperbolic tree and rubber-sheet navigation. For *separate* displays, we speculated that perhaps interactive brushing and linking alone may not be sufficient to resolve the view-coordination problem; we may also need to add landmarks to help users associate different views Lam [2008]. In general, spatial arrangement in multiple-level display design is still an open research question.

We believe part of the reason for these gaps in our knowledge is due to the prevalence of evaluations that focus at the interface level, and the difficulty in conducting analyses to summarize published evaluation results. Our experience led us to draft out two directions of future work. First is the need for empirical evaluation to focus more at the interface-factor level, rather than at the interface level. While useful as a benchmarking tool, evaluations that focus at the interface level cannot uncover interplays between interface factors. Given the diversity in design requirements, results from these studies are difficult to reapply. Our recommendations in methodology are published elsewhere Lam and Munzner [2008], echoing a previous call from Chen & Yu Chen and Yu [2000] who conducted a meta-analysis on information visualization systems.

A second line of future work is to carve out a methodology that is useful when synthesizing existing results. Even though the number of visualization evaluations are increasing, it is still difficult to examine them collectively to obtain a better snapshot of our knowledge in visualizations. Part of the challenge is to standardize empirical evaluations and reporting Chen and Yu [2000]. At the same time, we need to recognize the complexity in visualization evaluation and derive a methodology for reviewing existing results. Our effort here is a very primitive first effort, and we welcome discussions on the topic.

APPENDIX A

Reviewed Studies: Interfaces, Tasks, Data and Results

This appendix summarizes the key aspects of the multi-level interface studies reviewed in this paper: study interfaces, tasks, data, and statistically significant results for each study:

Interfaces: We classified study interfaces based on the taxonomy used in the article, as *loLevel (L)*, *temporal (T)*, *separate (S)*, or *embedded (E)*. We listed all the categories to which the interface were categorized. For example, a zoomable interface with an overview would be classified as "*separate + temporal*", or "*S+T*". We also included the names of the interfaces if they were provided in the original study papers.

Significant Results: We listed the statistically significant time and accuracy results, using the interface taxonomy of *L*, *T*, *S*, and *E*. Even though many studies reported questionnaires and observations, we do not include them due to space constraints.

A.1   [FCSCREEN] KEEPING THINGS IN CONTEXT: A COMPARATIVE EVALUATION OF FOCUS PLUS CONTEXT SCREENS, OVERVIEWS AND ZOOMING

This study compared three visualization techniques to extract information from large static documents and avoid collisions in a driving simulation Baudisch et al. [2002].

Interfaces:

A.1.1 f+c: focus plus context screens implemented as a wall-sized low-resolution displays with an embedded high-resolution display region, with panning interaction only (E)

A.1.2 o+d: overview plus detail with a high-level window + a smaller *temporal* screen (S+T)

A.1.3 z+p: traditional pan-and-zoom (T)

Task(s):

1. Static task: Find route in a map

2. Static task: Verify connection in a network

3. Dynamic task: Avoid collision in a computer-game like environment

Data:

- For static tasks: spatial map data

- For dynamic tasks: a computer-game like environment with a driving scene with falling objects. Some of which were visible at high levels (i.e., the rocks), and some only at low levels (i.e., the nails).

Significant Result(s):

Time:

- $E < T$ (*Find route; verify connection*)

- $E < [S + T]$ (*Find route; verify connection*)

    Accuracy:

- $E > [S + T]$(*Avoid collision*)

A.2   [FISHNET] FISHNET, A FISHEYE WEB BROWSER WITH SEARCH TERM POPOUTS: A COMPARATIVE EVALUATION WITH OVERVIEW AND LINEAR VIEW

A user study that helps practitioners determine which visualization technique – fisheye view, overview, or regular linear view – to pick for which type of visual search scenario in viewing web pages on browsers Baudisch et al. [2004].

Interfaces:

A.1.4 Fisheye: a non-scrollable browser with readable and non-readable texts, depending on the user selection (E)

A.1.5 Overview: loLevel plus a high-level view showing the entire web page fitted vertically to a fixed horizontal width (S)

A.1.6 Linear: a traditional browser interface with vertical scrolling (L)

Task(s):

1. *Outdated*: check if page contained all four search terms

2. *Product choice*: find cheapest notebook with four features

3. *Co-occurrence*: check if page contained any paragraphs that contained both search terms

4. *Analysis*: check how many times Mrs. Clinton was mentioned, with "Clinton" being the search term

Data: Web documents

Significant Result(s):
Time:

- $S < L$ (*Outdated*)

- $E < L$ (*Outdated, Product choice*)

- $E < S$ (*Product choice*)

- $L < E$ (*Co-occurrence*)

Accuracy:

- $E > L > S$ (*Co-occurrence*)

A.3 [DATELENS] DATELENS: A FISHEYE CALENDAR INTERFACE FOR PDAS

This study compared between two types of calendar visualizations: DateLens and Pocket PC 2002 calendar for both simple and complex tasks.

Interfaces:

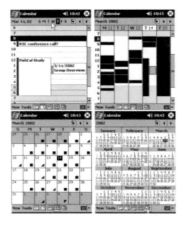
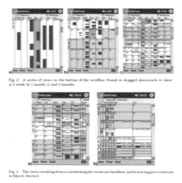

A.1.7 Pocket PC: default Pocket PC calendar, providing separate day, week, month and year views (T)

A.1.8 DateLens: a Table Lens-like distortion that can show multiple levels of details simultaneously, with the default configured as a 3-month view (E)

Task(s):

1. *Searching*: find the start and/or end dates of appointments

2. *Navigation and Counting*: navigate to particular appointments or monthly views, and count pre-defined activities

3. *Scheduling*: schedule an event of various time spans

Data: Calendar data

Significant Result(s):

Time:

- $T < E$(*Check schedule, Count Mondays/Sundays in a month, Find the closest free Saturday night/Sunday*)

- $E < T$(*Count conflicts/free days in a 3-month period, Find freest/busiest two-week period in the next three months, Find a start date for a specific activity, Find freest half-day in a month*)

Percent completed task:

- $E > T$, **except** for two tasks to find schedule details about specific activities

A.4   [SCATTERPLOT] USER INTERACTION WITH SCATTERPLOTS ON SMALL SCREENS—A COMPARATIVE EVALUATION OF GEOMETRIC-SEMANTIC ZOOM AND FISHEYE DISTORTION

This study compared between geometric plus semantic zooming and fisheye distortion to display scatterplots on PDA-sized screens for visual scan, information access, and visual comparison tasks Buring et al. [2006a].

Interfaces:

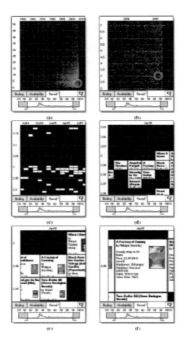

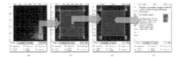

A.1.9 Geometric-semantic ZUI: geometric and semantic zooming of scatter plots (T) © 2006 IEEE

A.1.10 Fisheye distortion: rectangular fisheye with at most two zoom levels per display similar to the Table Lens implementation (E) © 2006 IEEE

Task(s):

Note: Only one example per task type was provided in the paper, and are listed here.

1. *Visual scan: How many books have been published since the year 2000 at a price of EUR 30?*

2. *Information access: Who is the author of the most expensive book published in the year 2005?*

3. *Comparison of information objects: Between August and November 2001 four books were published, which are available at a price of EUR 8.53. Which is the one with the most pages?*

Data: Two-dimensional scatter plot data (details not provided in paper)

Significant Result(s):
No significant objective performance results.

A.5   [ELIDESRC] HIDDEN MESSAGES: EVALUATING THE EFFICIENCY OF CODE ELISION IN PROGRAM NAVIGATION

This study examined using elision of program code in code navigation by studying a scrollable interface along with two elided interfaces Cockburn and Smith [2003].

Interfaces:

(a) Flat text.   (b) Legible elision.

A.1.11 Flat text: a non-eliding interface that showed program codes in legible, normal, font (L)

A.1.12 Legible elision: an eliding interface that showed elided text in a font just large enough to read (E1)

Task(s):

1. *Signature retrieval*: find argument type in methods

2. *Body retrieval*: find first call of a specific method in another method

3. *Combination of body search and signature retrieval*: find the return type of the method that is called last in another method

4. *Program browsing*: determine the longest method in the class

Data: Program code: small (160–200 lines) and large (360–400 lines)

Significant Result(s):
Time:

- $E2 < L$(*Signature retrieval, Combination of body search and signature retrieval*)

A.6 [FISHSTEER] FISHEYE VIEWS ARE GOOD FOR LARGE STEERING TASKS

This study tested the effects of magnification and representation on user performance in a basic pointing activity called steering – where a user moves a pointer along a predefined path in the workspace. Researchers tested three types of fisheye at several levels of distortion, and also compared the fisheyes with two non-distorting overview + detail techniques Gutwin and Skopik [2003].

Interfaces:

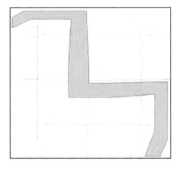
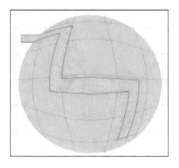
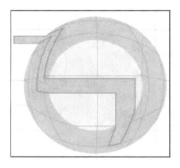

A.1.13 Sarkar-and-Brown fisheye (E1)   A.1.14 Round-lens fisheye (E2)   A.1.15 Flat-lens fisheye (E3)

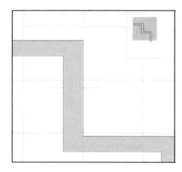

A.1.16 Panning view (S1)   A.1.17 Radar view (S2)

Task(s): 2D-steering task that required participants to move a pointer along a path that is defined by objects in a visual workspace. In order to perform the task, participants needed to use the low-level view for accurate steering and the high-level view to pan around.

Data: Abstract 2D paths: horizontal, diagonal, step, curve

Significant Result(s):
Time:

- $E \leq S$ (at all magnification levels)

Accuracy:

- $E \geq S$ (at all magnification levels)

A.7   [BIGONSMALL] INTERACTING WITH BIG INTERFACES ON SMALL SCREENS: A COMPARISON OF FISHEYE, ZOOM, AND PANNING TECHNIQUES

This study compared three techniques for using large interfaces on small screens – panning, Two-level zoom, and a fisheye view – using three common tasks Gutwin and Fedak [2004b].

Interfaces:

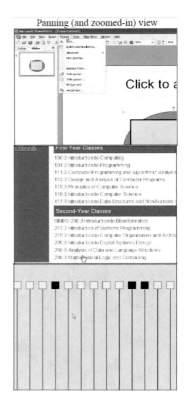 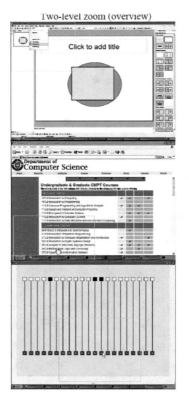 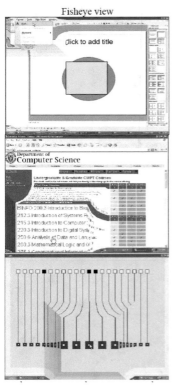

A.1.18 Panning: a "sliding window" interface that showed a portion of the source screen at full size (T)	A.1.19 Two-level zoom: a zoomable interface with an overview that showed the entire screen in a reduced form, and at full size when zoomed in (S)	A.1.20 Fisheye view: an interface that showed a magnified region implemented with a flat-top pyramid lens (E)

Tasks and Data:

1. *Editing*: create a presentation document and add objects to a slide

2. *Navigation*: visit a given sequence of links in a web page as quickly as possible

3. *Monitoring*: detect failures and restart failed devices in a simulated real-time device array

Significant Result(s):

Time:

- $E = S < T$ *(Editing)*

- $E < S = T$ *(Navigation)*

- $S < E < T$ *(Monitoring)*

A.8   [EDOC] READING OF ELECTRONIC DOCUMENTS: THE USABILITY OF LINEAR, FISHEYE AND OVERVIEW + DETAIL INTERFACES AND READING PATTERNS AND USABILITY IN VISUALIZATION OF ELECTRONIC DOCUMENTS

This study explored reading patterns and usability in visualizations of electronic documents using a fisheye, an overview + detail, and a linear interface with question answering and essay tasks. Hornbæk and Frokjær [2001], Hornbæk et al. [2003].

Interfaces:

Fig. 1(a). The linear interface.

Fig. 1(c). The overview+detail interface.

Fig. 1(b). The fisheye interface.

A.1.21 Linear: traditional vertically scrollable interface (L)

A.1.22 o+d: Overview+Detail: *loLevel* plus a high-level overview of the entire document, reduced by 1:17 in size on average, and coordinated with the high-level view. In the high-level view, only the section and subsection headers of the document were readable, with the rest of the document shrunk to fit within the available space (S)

A.1.23 Fisheye (Non-scrollable browser with only the most important part of the document was readable. The relative importance determined by the interface *a priori*. Participants could expand or collapse different parts of the documents by a mouse click (E)

Task(s):

1. *Essay*: read a document, from memory: (a) write 1-page essay, stating the main theses and ideas of the documents; (b) answer 6 incidental-learning questions

2. *Question-answering*: answer 6 questions

Data: Electronic text documents

Significant Result(s):
Time:

- $E < L$ (*Essay*)

- $E < S$ (*Essay*)

- $L < E$ (*Question-answering*)

Effectiveness:

- $S > L$ (*Essay: Author's grading*)

- $S > E$ (*Essay: Author's grading, Essay: # correct incidental-learning questions*)

- $L > E$ (*Essay: # correct incidental-learning questions*)

A.9   [ZUINAV] NAVIGATION PATTERNS AND USABILITY OF ZOOMABLE USER INTERFACES WITH AND WITHOUT AN OVERVIEW

This study compared zoomable user interfaces with and without an overview to understand the navigation patterns and usability of these interfaces using map data Hornbæk et al. [2002].

Interfaces:

A.1.24 Zoomable User Interface (ZUI): displayed a map, zoomable at 20 scale levels (T)

A.1.25 ZUI with Overview: *Temporal* plus a high-level view that was one-sixteenth the size of the zoomable window (S+T)

Task(s):

1. *Navigation*: find a well-defined map object

2. *Browsing*: scan a large area, possibly the entire map for objects of a certain type

3. *Label cities and counties*: write down as many objects within the a map area from memory

4. *Recognize cities*: circle all cities within a county and cross out cities that were believed to be outside of the county

Data: Geographical map:

• Washington map: 3 levels (county, city and landmark)

- Montana map: single level

Significant Result(s):
Time:

- $T < [S + T]$ (*Navigation*)

Accuracy:

- $T > [S + T]$ (Washington map: *Label cities and counties*, *Recognize cities*)

A.10   [FISHMENU] UNTANGLING THE USABILITY OF FISHEYE MENUS

This study investigated whether fisheye menus are useful as compared to the hierarchical menu and two variants of the fisheye menu, based on known-item search and browsing tasks Hornbæk and Hertzum [2007].

Task(s):

1. *Known-item search*

2. *Browsing*

Data:

- alphabetical data with 100 items
- categorical data with 292 items (3 levels of 4, 4x8, and 4x8x8 items)

Significant Result(s):
Time:

- T < all interfaces (*Known-item search*)

Accuracy:

- T > all interfaces (*Known-item search*)

Interfaces:

A.1.26 Hier-archical menu: traditional cas-cading menu. For the smaller data set, the menu had two levels. For the larger data set, the menu had three levels, or two submenus (T)

A.1.27 Overview: a high-level pane showing an index of letters of the items included in the menu and a low-level pane showing menu items. The portion of the items showed was determined by the mouse position relative to length of the menu (S)

A.1.28 Fisheye: a high-level pane showed an index of letters of the menu items and a low-level pane showed all the menu items, with a regular font-sized region surrounded by decreasing font sizes. At the two extreme ends, the items were unreadable (E+S)

A.1.29 Multi-focus: has two types of low-level regions: the mouse-selected menu items, and those that were determined to be significant based on *a priori* importance (E)

A.11 [FISHSRC] EVALUATING A FISHEYES OF SOURCE CODE

This study compared the usability of the fisheye view with a common, linear presentation of program source code Jakobsen and Hornbæk [2006].

Interfaces:

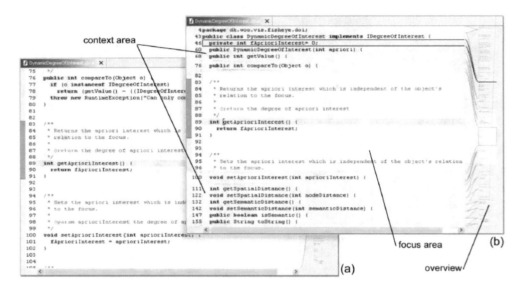

Figure 1. Screenshots of (a) the Linear and (b) the Fisheye interface showing the same source file of 161 lines.

A.1.30 Linear: vertically scrollable and displayed all the program lines (L) and Fisheye (No vertical scrolling, but selectively displaying semantically relevant parts of the source code based on the lines displayed in the focal region. The selection was determined by a modified version of Furnas' degree-of-interest function, where semantic distance was also considered along with syntactic distance and *a priori* significance (E)

Task(s):

1. *One-step navigation*

2. *Two-step navigation*

3. *Determine field encapsulation*

4. *Determine delocalization*

5. *Determine control structure*

Data: Program source code

Significant Result(s):
Time:

- $E < L$ (*Two-step navigation*: 15%, *Determine delocalization*: 30%)

A.12   [SUMTHUM] SUMMARY THUMBNAILS: READABLE OVERVIEWS FOR SMALL SCREEN WEB BROWSERS

The study compared Summary Thumbnails – thumbnail views enhanced with readable text fragments – with thumbnails, a single-column interface, and a desktop interface in a number of web information search tasks Lam and Baudisch [2005].

Interfaces:

A.1.31 Summary Thumbnail / Thumbnail: scaled-down image of the original web page fitted to the width of the PDA screen, with or without preserving the readability of the text) (T)

A.1.32 Desktop: Original, unscaled desktop-sized web page (L)

Task(s): Information searches

Data: Web documents

Significant Result(s): No significant differences in performance time or task accuracy

A.13 [LINEGR] OVERVIEW USE IN MULTIPLE VISUAL INFORMATION RESOLUTION INTERFACES

The study looked at overview use in two multi-level interfaces with high-level displays either embedded within, or separate from, the overviews using finding and matching tasks Lam et al. [2007].

Interfaces:

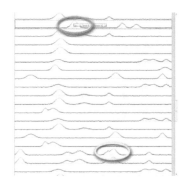
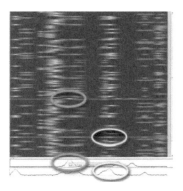
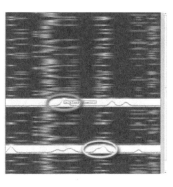

A.1.33 loLevel: stacked line graph plots, encoding the x and the y line graph values with space, and the y-values doubly encoded with colour (L) © 2007 IEEE

A.1.34 separate: low-level interface with strips that encode the y-values of the line graph data with colour alone. Mouse-click on strip displays high-level plots in a separate panel (S) © 2007 IEEE

A.1.35 embedded: low-level regions of lam07. Mouse-click on strip displays high-level plots in place (E) © 2007 IEEE

Task(s):

1. *Find highest point*

2. *Find most number of peaks in line graph*

3. *Match a small region of line graph*

4. *Match entire line graph*

Data: 140 line graphs, each with 800 data points

Significant Result(s):
Time:

- $S < L$ (*Find highest point*)

- $E < L$ (*Find highest point*)

A.14 [RUBNAV] AN EVALUATION OF PAN AND ZOOM AND RUBBER SHEET NAVIGATION

This study evaluated two navigation techniques with and without an overview. The techniques examined are conventional pan and zoom navigation and rubber sheet navigation, a rectilinear focus + context technique Nekrasovski et al. [2006].

Interfaces:

A.1.36 PNZ: the traditional pan and zoom interface augmented with a visual cue to indicate the location of the target branch as colouring of the node regardless of the allotted screen presence (T)

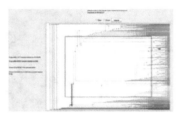

A.1.37 RSN: implemented the Rubber Sheet Navigation, augmented with a Halo-like arc served as the visual cue, as the actual target may be off screen (E)

A.1.38 PNZ+OV: add high-level overview in addition to the *temporal* view (T+S)

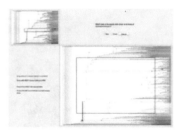

A.1.39 RNS+OV: add high-level overview in addition to the *embedded* view(E+S)

Task(s): Compare the topological distances between coloured nodes in a large tree and determine which of the distances was smaller

Data: Large trees

Significant Result(s):
Time:

- $T < E$

- $[T + S] < [E + S]$

A.15 [SNAP] SNAP-TOGETHER VISUALIZATION: CAN USERS CONSTRUCT AND OPERATE COORDINATED VISUALIZATIONS

This study explored coordination construction and operation in Snap-together visualization operating an overview-and-detail coordination, a detail-only and an uncoordinated interface to display census data North and Shneiderman [2000].

Interfaces:

not available

A.1.40 Detail-only: displayed census information grouped by geographic states (L)

A.1.41 Coordination / no-coordination (*loLevel* plus a high-level pane that displayed an alphabetical list of states included in the census (S)

Task(s):

1. *Coverage*: answer present or absent of objects

2. *Overview patterns*

3. *Visual / nominal lookup*

4. *Compare two or five items*

5. *Search for target value*

6. *Scan all*

Data: United States census data

Significant Result(s):
Time:

- $S(\pm coord) < L$ (*Coverage, Overview patterns*)

- $S(+coord) < L|S(-coord)$ (*Nominal lookup, Compare, Search, Scan*)

A.16   [INFOSCENT] THE EFFECTS OF INFORMATION SCENT ON VISUAL SEARCH IN THE HYPERBOLIC TREE BROWSER

The paper presents two experiments that investigated the effect of information scent (tasks with different Accuracy of Scent scores) on performance with the hyperbolic tree browser and the Microsoft Windows file browser Pirolli et al. [2003].

Interfaces:

A.1.42 Microsoft File Browser (T)   A.1.43 Hyperbolic tree browser
© 2003 IEEE   (E) © 2003 IEEE

Task(s):

1. *Information Retrieval*: simple, complex

2. *Comparison*: local, global

Data: CHI'97 BrowseOff tree, trimmed to four levels with 1436 nodes, and 66 nodes at the lowest level

Significant Result(s):
Time:

- $E < T$ (High-scent tasks)
- $T < E$ (Low-scent tasks)

A.17 [SPACETREE] SPACETREE: SUPPORTING EXPLORATION IN LARGE NODE LINK TREE, DESIGN EVOLUTION AND EMPIRICAL EVALUATION

The study compared SpaceTree – a novel tree browser with dynamic rescaling of branches of the tree – with the hyperbolic tree browser and the Windows explorer in a series of locate, refind, and topology-related tasks Plaisant et al. [2002].

Interfaces:

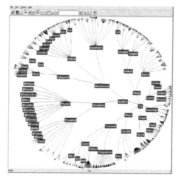
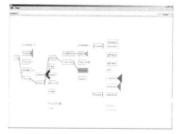

A.1.44 Microsoft Explorer file browser (T) © 2002 IEEE

A.1.45 Hyperbolic tree browser: lays out a tree based on a non-Euclidean hyperbolic plane ($E_{hyperbolic}$) © 2002 IEEE

A.1.46 SpaceTree: dynamically rescales the tree branches for the available screen space, preserves ancestral nodes but elides the rest into a triangular icon ($E_{spaceTree}$) © 2002 IEEE

Task(s):

1. *Node searches*

2. *Search of previously visited nodes*

3. *Topology questions*

Data: CHI'97 BrowseOff tree with over 7000 nodes

Significant Result(s):
Time:

- $T < E_{hyperbolic}$ (*Node searches*: 1 out of 3 tasks)

- $E_{spaceTree} < T$ (*Node searches*: 1 out of 3 tasks)

- $T < E_{hyperbolic}$ (*Refind previously visited nodes*)

- $E_{spaceTree} < E_{hyperbolic}$ (*Refind of previously visited nodes*)

- $T < E_{spaceTree}$ (*Refind of previously visited nodes*)

- $E_{spaceTree} < T$ (*Topology*: list all ancestor nodes)

- $E_{hyperbolic} < E_{spaceTree}$ (*Topology*: local topology)

Accuracy:

- $E_{spaceTree} > E_{hyperbolic} > T$ (*Refind of previously visited nodes, Topology*: overview)

A.18 [VISMEM] ZOOMING, MULTIPLE WINDOWS, AND VISUAL WORKING MEMORY

The paper presents a theoretical model of performance that models the relative benefits of these techniques when used by humans for completing a task involving comparisons between widely separated groups of objects based on a user study of zooming and multiple windows interfaces Plumlee and Ware [2006].

Interfaces:

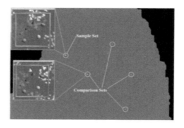

A.1.47 Zooming: continuous zoom mechanism (T)

A.1.48 Multiple Windows: Two levels: up to two low-level windows selected from a high-level view. The targets were clusters of 3-D geometric objects. Their high-level view showed only the location of the candidate targets, but not the details. At the intermediate levels, the target locations and details were camouflaged by the textured background. The lowest level presented enough target details for the visual comparison (S)

Task(s): Multiscale comparison task to find a cluster that matched the sample set of 3D objects.

Data: Six targets, each a cluster of 3D geometric objects with 1 to 7 items, each item taken from five possible shapes

Significant Result(s):
Time:

- $T < S$ (sets with one or two items)

- $S < T$ (sets with five or seven items)

Accuracy:

- $S < T$

A.19 [TIMEGR] VISUALIZATION OF GRAPHS WITH ASSOCIATED TIMESERIES DATA

This study evaluated and ranked graph+timeseries visualization options based on users' performance time and accuracy of responses on predefined tasks Saraiya et al. [2005].

Interfaces:

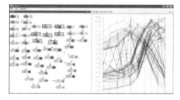

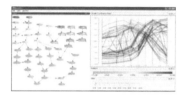

A.1.49 Multiple-Attribute Single-View (MS): displayed all 10 time points simultaneously as simple glyphs, representing the nodes of the graph (L) © 2005 IEEE

A.1.50 Single-Attribute Single-View (SS): displayed the value of the time points as colour of the nodes, linked with a user-controlled slider bar to view the other nine time points (T) © 2005 IEEE

Task(s):

1 time point:

- Read value, search node

2 time points:

- Determine change in values

10 time points:

- Determine time trend, topology trend

- Search time point, search trend

- Identify a outlier group

Data: 50-node graph, each node showing a timeseries with 10 time points

Significant Result(s):

Time:

- $T < L$ (*Topology trend,*)

- $L < T$ (*Outlier, Search time point*)

Accuracy:

- $T \geq L$ (all tasks **except** *Outlier*)
- $L > T$ (*Outlier*)

A.20 [FISHRADAR] A COMPARISON OF TRADITIONAL AND FISHEYE RADAR VIEW TECHNIQUES FOR SPATIAL COLLABORATION

This study compared an enhanced design that uses fisheye techniques with a traditional approach to radar views in spatial collaboration activities Schafer and Bowman [2003].

Interfaces:

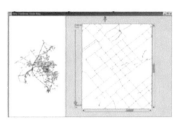

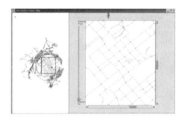

A.1.51 Traditional: contained a high-level view linked to a low-level view (S)

A.1.52 Fisheye: fisheye high-level view coupled with a low-level view (E)

Task(s): Collaborative traffic and road-sign positioning. 2 participants, each with partial information to position signs

Data: Map

Significant Result(s): Participants required less verbal communications with *E* than *S*

A.21 [FISHNAV] NAVIGATING HIERARCHICALLY CLUSTERED NETWORKS THROUGH FISHEYE AND FULL-ZOOM METHODS

This experiment compared two methods for viewing hierarchically clustered networks: the traditional full-zoom techniques provide details of only the current level of the hierarchy; and the fisheye views, generated by the variable-zoom algorithm, to provide information about higher levels as well Schaffer et al. [1996].

Task(s): Find and repair a broken telephone line in the network by rerouting a connection between two endpoints of the network that contained the break

Data: Hierarchical network of 154 nodes with 39 clusters

Significant Result(s):
Time:

- $E < T$ (*Repair*)

Interfaces:

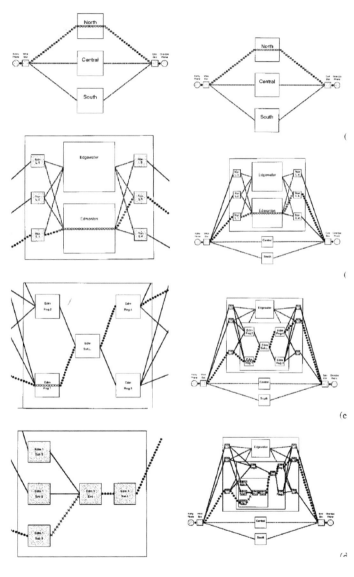

A.1.53 Full-Zoom: displayed children nodes of a single parent at the same level (T)

A.1.54 Fisheye: displayed the same children nodes along with all the ancestral nodes acting as context (E)

A.22 [SPACEFILL] AN EVALUATION OF CONTENT BROWSING TECHNIQUES FOR HIERARCHICAL SPACE-FILLING VISUALIZATIONS

The paper presents two experiments that compared a distortion algorithm based on fisheye and continuous zooming techniques with the drill-down method for browsing data in treemaps with or without the need for context Shi et al. [2005].

Interfaces:

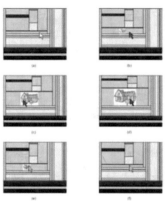

Figure 4: Implementation results of the algorithm described above. The red arrow indicates selection (b)-(d) and the green arrow indicates release of the mouse button (e) and (f).

A.1.55 Drill-Down: traditional TreeMap display, where the display showed only nodes from the same level of the same branch of the tree (T)

A.1.56 Distortion: retained all the ancestral levels of the displayed nodes, using distortion to fit all the nodes in the display (E) © 2005 IEEE

Task(s):

1. *Browsing*: find an image

2. *Browsing with Context*: find target based on its neighboring images and their interrelations, or context, defined as "a set of images spatially and hierarchically related in a certain configuration" (p. 86). This context was held constant for all trials, and involved multiple levels of the tree

Data:

2 hierarchies, both had 30 different images and >300 files of other formats

• Deep: 6 levels, <=3 subdirectories/level

- Wide: 3 levels, $<=6$ subdirectories/level

Significant Result(s):
Time:

- $E < T$ (*Browsing*: 65% faster with wide, 156% faster with deep; *Browsing with Context*: 61% faster with wide, 84% faster with deep)

Effectiveness:

- $E > T$ (gave up)

- $T > E$ (timed out)

Bibliography

Baudisch, P., Good, N., Bellotti, V., and Schraedley, P. 2002. Keeping Things in Context: A Comparative Evaluation of Focus Plus Context Screens, Overviews, and Zooming. In *Proc. ACM SIGCHI Conf. on Human Factors in Computing Systems (CHI'02)*. 259–266. DOI: 10.1145/503376.503423 14, 16, 22, 23, 24, 31, 36, 37, 41, 43, 44, 45, 54

Baudisch, P., Lee, B., and Hanna, L. 2004. Fishnet, a Fisheye Web Browser with Search Term Popouts: A Comparative Evaluation with Overview and Linear View. In *Proc. ACM Advanced Visual Interfaces (AVI'04)*. 133–140. DOI: 10.1145/989863.989883 13, 14, 18, 19, 24, 25, 26, 27, 28, 29, 30, 41, 42, 45, 56

Bederson, B. 2000. Fisheye Menus. In *Proc. ACM SIGCHI Symposium on User Interface Software and Technology (UIST'00)*. 217–226. DOI: 10.1145/354401.354782 16, 24

Bederson, B., Clamage, A., Czerwinski, M., and Robertson, G. 2004. DateLens: A Fisheye Calendar Interface for PDAs. *ACM Trans. on Computer-Human Interaction (TOCHI) 11*, 1 (Mar.), 90–119. DOI: 10.1145/972648.972652 13, 14, 24, 27, 36, 37, 38, 45

Bederson, B., Shneiderman, B., and Wattenberg, M. 2002. Ordered and Quantum Treemaps: Making Effective Use of 2D Space to Display Hierarchies. *ACM Transactions on Graphics (TOG) 21*, 4, 833–854. DOI: 10.1145/571647.571649 5

Buring, T., Gerken, J., and Reiterer, H. 2006a. Usability of Overview-Supported Zooming on Small Screens with Regard to Individual Differences in Spatial Ability. In *Proc. ACM AVI Workshop on BEyond time and errors: novel evaLuation methods for Information Visualization*. 223–240. DOI: 10.1145/1133265.1133310 36, 60

Buring, T., Gerken, J., and Reiterer, H. 2006b. User Interaction with Scatterplots On Small Screens— A Comparative Evaluation of Geometric-Semantic Zoom and Fisheye Distortion. *IEEE Trans. on Visualization and Computer Graphics (TVCG) 12*, 5, 829–836. DOI: 10.1109/TVCG.2006.187 13, 14

Card, S., Mackinlay, J., and Shneiderman, B. 1999. *Readings in Information Visualization: Using Vision to Think*. Morgan Kaufmann, San Francisco, California. 1, 15, 31, 32, 35, 41, 44

Carpendale, M., Cowperthwaite, D., and Fracchia, F. 1997. Making Distortions Comprehensible. In *Proc. IEEE Symposium on Visual Languages*. 36–45. 44

Chen, C. and Yu, Y. 2000. Empirical Studies of Information Visualization: A Meta-Analysis. *Intl. J. of Human-Computer Studies (IJHCS) 53*, 851–866. DOI: 10.1006/ijhc.2000.0422 9, 51

Cockburn, A., Karlson, A., and Bederson, B. 2008. A Review of Overview+Detail, Zooming, and Focus+Context Interfaces. *ACM Computing Surveys 41*, 1. DOI: 10.1145/1456650.1456652 1, 4, 7, 12

Cockburn, A. and Smith, M. 2003. Hidden Messages: Evaluating the Effectiveness of Code Elision in Program Navigation. *15*, 3, 387–407. 14, 15, 18, 19, 24, 26, 27, 28, 45, 62

Ellis, G. and Dix, A. 2006. An Exploratory Analysis of User Evaluation Studies in Information Visualization. In *Proc. ACM AVI Workshop on BEyond time and errors: novel evaLuation methods for Information Visualization.* 1–7. DOI: 10.1145/1168149.1168152 9

Ellis, G. and Dix, A. 2007. A Taxonomy of Clutter Reduction for Information Visualization. *IEEE Trans. on Visualization and Computer Graphics (TVCG) 13*, 6, 1216–1223. DOI: 10.1109/TVCG.2007.70535 3

Furnas, G. 1986. Generalized Fisheye Views. In *Proc. ACM SIGCHI Conf. on Human Factors in Computing Systems (CHI'86).* 16–23. DOI: 10.1145/22339.22342 3, 29, 44

Furnas, G. W. 2006. A Fisheye Follow-Up: Further Reflection on Focus + Context. In *Proc. ACM SIGCHI Conf. on Human Factors in Computing Systems (CHI'06).* 999–1008. DOI: 10.1145/1124772.1124921 1, 3, 31, 32, 35, 41, 49

Gutwin, C. and Fedak, C. 2004a. A Comparison of Fisheye Lenses for Interactive Layout Tasks. In *Proc. Conf. on Graphics Interface (GI'04).* 213–220. 41, 42, 43, 45

Gutwin, C. and Fedak, C. 2004b. Interacting with Big Interfaces on Small Screens: a Comparison of Fisheye, Zoom, and Panning Techniques. In *Proc. Conf. on Graphics Interface (GI'04).* 145–152. 14, 18, 20, 24, 26, 27, 28, 36, 38, 66

Gutwin, C. and Skopik, A. 2003. Fisheyes are Good for Large Steering Tasks. In *Proc. ACM SIGCHI Conf. on Human Factors in Computing Systems (CHI'03).* 201–208. 13, 14, 24, 45, 64

Herman, I., Melancon, G., and Marshall, M. 2000. Graph Visualization and Navigation in Information Visualization: A Survey. *IEEE Trans. on Visualization and Computer Graphics (TVCG) 6*, 1, 24–43. DOI: 10.1109/2945.841119 35

Hornbæk, K., Bederson, B., and Plaisant, C. 2002. Navigation Patterns and Usability of Zoomable User Interfaces with and without an Overview. *ACM Trans. on Computer-Human Interaction (TOCHI) 9*, 4, 362–389. DOI: 10.1145/586081.586086 13, 14, 15, 16, 17, 18, 19, 22, 23, 24, 28, 29, 32, 36, 70

Hornbæk, K. and Frokjær, E. 2001. Reading of Electronic Documents: The Usability of Linear, Fisheye and Overview+Detail Interfaces. In *Proc. ACM SIGCHI Conf. on Human Factors in Computing Systems (CHI'01)*. 293–300. DOI: 10.1145/365024.365118 5, 7, 13, 14, 16, 17, 18, 20, 24, 26, 27, 29, 30, 31, 32, 41, 42, 45, 68

Hornbæk, K., Frokjaer, E., and Plaisant, C. 2003. Reading Patterns and Usability in Visualization of Electronic Documents. *ACM Trans. on Computer-Human Interaction (TOCHI) 10*, 2, 119–149. DOI: 10.1145/772047.772050 13, 14, 16, 17, 18, 20, 24, 26, 27, 29, 30, 31, 32, 41, 42, 68

Hornbæk, K. and Hertzum, M. 2007. Untangling the Usability of Fisheye Menus. *ACM Trans. on Computer-Human Interaction (TOCHI) 14*, 2. DOI: 10.1145/1275511.1275512 14, 15, 16, 22, 24, 25, 27, 28, 36, 41, 45, 50, 72

Hudhausen, C., Douglas, S., and Stasko, J. 2002. A Meta-Study of Algorithm Visualization Effectiveness. *Journal of Visual Languages and Computing 13*, 259–290. DOI: 10.1006/jvlc.2002.0237 9

Jakobsen, M. R. and Hornbæk, K. 2006. Evaluating A Fisheye View of Source Code. In *Proc. ACM SIGCHI Conf. on Human Factors in Computing Systems (CHI'06)*. 377–386. DOI: 10.1145/1124772.1124830 14, 16, 26, 27, 28, 29, 30, 31, 45, 50, 74

Jul, S. and Furnas, G. 1998. Critical Zones in Desert Fog: Aids to Multiscale Navigation. In *Proc. ACM SIGCHI Symposium on User Interface Software and Technology (UIST'98)*. 97–106. DOI: 10.1145/288392.288578 32

Keim, D., Mansmann, F., Schneidewind, J., and Ziegler, H. 2006. Challenges in Visual Data Analysis. In *Proc. IEEE Information Visualization (IV'06)*. 9–16. DOI: 10.1109/IV.2006.31 3

Lam, H. 2008. A Framework of Interaction Costs in Information Visualization. *IEEE Transactions on Visualization and Computer Graphics (TVCG) 14*, 6, 1149–1156. DOI: 10.1109/TVCG.2008.109 15, 51

Lam, H. and Baudisch, P. 2005. Summary Thumbnails: Readable Overviews for Small Screen Web Browsers. In *Proc. ACM SIGCHI Conf. on Human Factors in Computing Systems (CHI'05)*. 681–690. DOI: 10.1145/1054972.1055066 14, 18, 26, 27, 28, 76

Lam, H. and Munzner, T. 2008. Increasing the Utility of Quantitative Empirical Studies for Meta-analysis. In *Proc. CHI Workshop on BEyond time and errors: novel evaLuation methods for Information Visualization (BELIV'08)*. 21–27. DOI: 10.1145/1377966.1377969 9, 48, 51

Lam, H., Munzner, T., and Kincaid, R. 2007. Overview Use in Multiple Visual Information Resolution Interfaces. *IEEE Transactions on Visualization and Computer Graphics (TVCG) 13*, 6, 1278–1283. DOI: 10.1109/TVCG.2007.70583 5, 13, 14, 16, 17, 18, 19, 24, 26, 27, 28, 29, 41, 44, 45, 50, 77

Lam, H., Rensink, R., and Munzner, T. 2006. Effects of 2D Geometric Transformations on Visual Memory. In *Proc. Symposium on Applied Perception in Graphics and Visualization (APGV'06)*. 119–126. DOI: 10.1145/1140491.1140515 44

Lamping, J., Rao, R., and Pirolli, P. 1995. A Focus+Context Technique Based On Hyperbolic Geometry for Visualizing Large Hierarchies. In *Proc. ACM SIGCHI Conf. on Human Factors in Computing Systems (CHI'95)*. 401–408. DOI: 10.1145/223904.223956 44

Lau, K., Rensink, R., and Munzner, T. 2004. Perceptual Invariance of Nonlinear Focus+Context Transformations. In *Proc. ACM Symposium on Applied Perception in Graphics and Visualization (APGV'06)*. 65–72. DOI: 10.1145/1012551.1012563 44

Leung, Y. and Apperley, M. 1994. A Review and Taxonomy of Distortion-Oriented Presentation Techniques. *ACM Trans. on Computer-Human Interaction (TOCHI) 1*, 2 (June), 126–160. DOI: 10.1145/180171.180173 12

Mountjoy, D. 2001. Perception-Based Development and Performance Testing of a Non-linear Map Display. Ph.D. thesis, North Carolina State University, North Carolina, USA. 44

Nekrasovski, D., Bodnar, A., McGrenere, J., Munzner, T., and Guimbretière, F. 2006. An Evaluation of Pan and Zoom and Rubber Sheet Navigation. In *Proc. ACM SIGCHI Conf. on Human Factors in Computing Systems (CHI'06)*. 11–20. DOI: 10.1145/1124772.1124775 14, 22, 24, 36, 38, 41, 42, 45, 46, 78

Nigay, L. and Vernier, F. 1998. Design Method of Interaction Techniques for Large Information Spaces. In *Proc. of Advanced Visual Interfaces (AVI'98)*. 37–46. DOI: 10.1145/948496.948503 32

Norman, D. 1993. *Things That Make Us Smart: Defending Human Attributes in the Age of the Machine*. Perseus Books. 49, 50

North, C. and Shneiderman, B. 2000. Snap-Together Visualization: Can Users Construct and Operate Coordinated Visualizations. *Intl. J. of Human-Computer Studies (IJHCS) 53*, 5, 715–739. DOI: 10.1006/ijhc.2000.0418 14, 26, 27, 31, 42, 80

Pirolli, P. 2007. *Information Foraging Theory: Adaptive Interaction with Information*. Oxford University Press, USA. DOI: 10.1093/acprof:oso/9780195173321.001.0001 48

Pirolli, P., Card, S., and van der Wege, M. 2003. The Effects of Information Scent on Visual Search in the Hyperbolic Tree Browswer. *ACM Trans. on Computer-Human Interaction (TOCHI) 10*, 1, 20–53. DOI: 10.1145/606658.606660 10, 13, 14, 16, 24, 25, 36, 37, 41, 42, 45, 46, 82

Plaisant, C. 2004. The Challenge of Information Visualization Evaluation. In *Proc. ACM Advanced Visual Interfaces (AVI'04)*. 109–116. DOI: 10.1145/989863.989880 9

Plaisant, C., Carr, D., and Shneiderman, B. 1995. Image-Browser Taxonomy and Guidelines for Designers. *IEEE Software 12*, 2, 21–32. DOI: 10.1109/52.368260 5

Plaisant, C., Grosjean, J., and Bederson, B. 2002. SpaceTree: Supporting Exploration in Large Node Link Tree, Design Evolution and Empirical Evaluation. In *Proc. IEEE Symposium on Information Visualization (InfoVis'02)*. 57–64. DOI: 10.1109/INFVIS.2002.1173148 14, 24, 36, 37, 42, 45, 46, 83

Plumlee, M. and Ware, C. 2006. Zooming versus Multiple Window Interfaces: Cognitive Costs of Visual Comparisons. *Proc. ACM Transactions on Computer-Human Interaction (TOCHI) 13*, 2, 179–209. DOI: 10.1145/1165734.1165736 13, 14, 23, 24, 36, 38, 42, 85

Saraiya, P., Lee, P., and North, C. 2005. Visualization of Graphs with Associated Time-series Data. In *Proc. IEEE Symposium on Information Visualization (InfoVis'05)*. 225–232. DOI: 10.1109/INFVIS.2005.1532151 13, 14, 23, 26, 32, 87

Sarkar, M., Snibbee, S., Tversky, O., and Reiss, S. 2003. Stretching the Rubber Sheet: A Metaphor for Viewing Large Layouts on Small Screens. In *Proc. ACM SIGCHI Symposium on User Interface Software and Technology (UIST'89)*. 81–91. DOI: 10.1145/168642.168650 46

Schafer, W. and Bowman, D. A. 2003. A Comparison of Traditional and Fisheye Radar View Techniques for Spatial Collaboration. In *Proc. ACM Conf. on Graphics Interface (GI'03)*. 39–46. 14, 24, 41, 45, 46, 89

Schaffer, D., Zuo, Z., Greenberg, S., Bartram, L., Dill, J., Dubs, S., and Roseman, M. 1996. Navigating Hierarchically Clustered Networks through Fisheye and Full-Zoom Methods. *ACM Trans. on Computer-Human Interaction (TOCHI) 3*, 2 (Mar.), 162–188. DOI: 10.1145/230562.230577 14, 24, 36, 37, 38, 45, 90

Shi, K., Irani, P., and Li, B. 2005. An Evaluation of Content Browsing Techniques for Hierarchical Space-Filling Visualizations. In *Proc. IEEE Symposium on Information Visualization (InfoVis'05)*. 81–88. DOI: 10.1109/INFOVIS.2005.4 14, 24, 36, 39, 42, 45, 92

Skopik, A. and Gutwin, C. 2003. Finding Things In Fisheyes: Memorability in Distorted Spaces. In *Proc. Conf. on Graphics Interface (GI'03)*. 47–56. 44

Skopik, A. and Gutwin, C. 2005. Finding Things In Fisheyes: Memorability in Distorted Spaces. In *Proc. ACM SIGCHI Conf. on Human Factors in Computing Systems (CHI'05)*. 771–780. 44

Thomas, J. and Cook, K. A., Eds. 2005. *Illuminating the Path: The Research and Development Agenda for Visual Analytics*. IEEE Computer Society Press. 41

Tullis, T. 1985. A Computer-Based Tool for Evaluating Alphanumeric Displays. In *Human-Computer Interaction: INTERACT '84*, B. Shackel, Ed. Elsevier Science, 719–723. 50

Valiati, E., Pimenta, M., and Freitas, C. 2006. A Taxonomy of Tasks for Guiding the Evaluation of Multidimensional Visualizations. In *Proc. ACM AVI Workshop on BEyond time and errors: novel evaLuation methods for Information Visualization.* 1–6. DOI: 10.1145/1168149.1168169 9

Winckler, M., Palanque, P., and Freitas, C. 2004. Tasks and Scenario-based Evaluation of Information Visualization Techniques. In *Proc. ACM Conf. on Task Models and Diagrams (TAMODIA'04).* 165–172. DOI: 10.1145/1045446.1045475 9

Zanella, A., Carpendale, M., and Rounding, M. 2002. On the Effects of Viewing Cues in Comprehending Distortions. In *Proc. ACM SIGCHI Conf. on Human Factors in Computing Systems (CHI'02).* 119–128. DOI: 10.1145/572020.572035 44

Authors' Biographies

HEIDI LAM

Heidi Lam has been a software engineer at Google, Inc. since 2008. She received a PhD from the University of British Columbia in 2008, studying with Dr. Munzner. She holds a master's in applied science from Simon Fraser University from 2004, and a bachelor's in biomedical engineering from SFU in 2002. She has also worked at Agilent Labs, Microsoft Research, Neoteric, and Nortel. Dr. Lam's current research interests include understanding how visualization can be used to support exploratory data analysis. She was co-chair of the ACM CHI 2010 workshop BELIV'10: BEyond time and errors: novel evaLuation methods for Information Visualization.

TAMARA MUNZNER

Tamara Munzner is associate professor at the University of British Columbia Department of Computer Science. She was a research scientist at the Compaq Systems Research Center (SRC), and earned her PhD from Stanford in 2000. Earlier she was a technical staff member at the National Science Foundation Research Center for Computation and Visualization of Geometric Structures (The Geometry Center) at the University of Minnesota from 1991 to 1995.

Her research interests include the development, evaluation, and characterization of information visualization systems and techniques from both user-driven and technique-driven perspectives. She has worked on visualization projects in a broad range of application domains, including evolutionary biology, genomics, systems biology, large-scale system administration, computer networking, web log analysis, computational linguistics, and geometric topology. She has consulted for or collaborated with many companies including Agilent, AT&T Labs, Google, Lytix, Microsoft, and Silicon Graphics.

Dr. Munzner was the IEEE Symposium on Information Visualization (InfoVis) Program/Papers Co-Chair in 2003 and 2004, and the Eurographics/IEEE Symposium on Visualization (EuroVis) Program/Papers Co-Chair in 2009 and 2010. She was a Member At Large of the Executive Committee of the IEEE Visualization and Graphics Technical Committee (VGTC) from 2004 through 2009, and she is currently a member of the InfoVis Steering Committee. She was one of the six authors of the 2006 Visualization Challenges Research report, commissioned by several directorates of the US National Science Foundation and National Institutes of Health.